Waynemiller

ELECTRICAL ENGINEERING

LICENSE REVIEW

Fourth Edition

Lincoln D. Jones

Donald G. Newnan

San Jose State University

REVIEW FOR THE NATIONAL ELECTRICAL

ENGINEERING EXAM USED BY 44 STATES

LIBRARY OF CONGRESS CATALOG CARD NUMBER 74-84808

INTERNATIONAL STANDARD BOOK NUMBERS

0-910554-17-X Paper 0-910554-18-8 Cloth

DISTRIBUTED BY

ENGINEERING PRESS P.O. BOX 5 SAN JOSE, CA. 95103

ENGINEERING CONSULTANTS P.O. BOX 6701 SAN JOSE, CA. 95150

NACSCORP (To Member Stores) 55 EAST COLLEGE ST. OBERLIN, OHIO 44074

MEDICAL & TECHNICAL BOOKS 2330 BROADWAY SANTA MONICA, CA. 90404

BRO-DART Stacey's 15255 EAST DON JULIAN RD. CITY OF INDUSTRY, CA. 91747

PRINTED IN U.S.A.

PREFACE

This edition reflects the authors evaluation of the kinds and distribution of problems that have appeared on past National Electrical Engineering Exams (also called the Principles and Practice of Electrical Engineering Exam) prepared by the National Council of Engineering Examiners (NCEE). The examination is described in detail in Chapter 1 - Introduction.

The 3rd edition of this book is the principal source of material for this new edition. One chapter (Profession Competence & Codes) was deleted and a new chapter (Typical Examination Set) was added. This new chapter is based on the authors evaluation of the most recent National Electrical Engineering exam and the review (see Appendix) of past examinations. The NCEE has a policy of denying permission to others to reproduce their exam problems. None of the problems in this book, therefore, are from their exams.

Although digital logic has not been included in past National E.E. exams as such, it is probable that one might expect questions of this nature in the future. For this reason the digital logic portion of the "Other Problems" chapter has been strengthened. Finally, the "Economic Analysis" chapter has been rewritten and expanded.

Every effort has been made to give the proper and correct solutions. There may be errors in the book, partly due to different possible interpretations. Any corrections or errors found would be gratefully received. Many people have aided and contributed to the preparation of the book. We are particularly grateful to Professors Jack Peterson (Univ. of Santa Clara), Howard Plotkin (Univ. of Houston), Michael O'Flynn and Rangaiya Rao (San Jose State Univ.), and to Zoltan Dios.

LDJ
DGN

CONTENTS

Engineer · In · Training License Review

By C. Dean Newnan

A review for the National Engineering Fundamentals Examination used by 47 States. For a detailed description of the current edition and for the price, write to:

Engineering Consultants
P.O. Box 6701
San Jose, California 95150

Engineering Fundamentals:

Principles, Problems, and Solutions

By Donald G. Newnan
and Bruce E. Larock

Engineering fundamentals are presented in ten subject areas: mathematics, statics, dynamics, mechanics of materials, fluid mechanics, thermodynamics, chemistry, electricity, economic analysis, and other problems. Each chapter begins with a review of important basic concepts, and continues with examples of the applications of these concepts. Next, problems are presented with detailed step-by-step solutions. Almost 700 actual exam problems with complete solutions are in the book.
Published by Wiley-Interscience. 592 pages. Clothbound.

A copy of the book may be obtained from your bookseller or from

Engineering Press
P.O. Box 5
San Jose, California 95103

Send $17.20 [Book $16.50; Postage & handling 70¢]
In California add Sales Tax.
Payment must accompany order.
The book price is set by Wiley-Interscience and is subject to change.

1

INTRODUCTION

This book is for electrical engineers seeking professional registration. The usual route to registration is by passing in turn two examinations. The first is the 8-hour Engineer-In-Training (or Fundamentals of Engineering) Exam. The emphasis here is on the broad range of engineering fundamentals common to all branches of engineering. The second exam in the 8-hour Professional Engineering (P.E.) Examination covering the principles and practice of electrical engineering.

In the past the various States prepared their own examinations. The trend, however, has been toward national examinations. At the present time the vast majority of the States have adopted the examinations prepared by the National Council of Engineering Examiners (NCEE). The examinations are given twice a year (April and November) simultaneous throughout the country.

The Engineer-In-Training (or Fundamentals of Engineering) Exam is the subject of other books and is not covered here. This book is concerned exclusively with the National Professional Engineering (Principles and Practices) Exam for electrical engineers. This national exam, prepared by NCEE, has been adopted by 44 of the 55 States and Other Jurisdictions of the U.S. that belong to NCEE.

The 8-hour examination is given in a one-day, two-part (morning and afternoon sessions) format. Each session contains 10 questions from which the examinee may select to solve any four questions. Each question is 10 points, making the maximum score 40 points per session. The questions are chosen from the following areas:

Area	Problems
Power & Systems	3
Machines	2
Electronics	2
Communications	2
Circuits	3
Controls	2
Economics	2
Instrumentation	2
Illumination	2
Total:	20 Problems

The list reflects the way NCEE characterizes the problems and their expected distribution. The examinations do not always follow the above "mix."

The authors have attempted to categorize examination problems within the normally accepted meanings and interpretations (see Appendix). The classification was done as part of the planning for this edition, and particularly for Chapter 11 - Typical Examination Set.

As in any examination, one wants the maximum possible points. The examiner that assigns these points must make his judgment based only on what is written down. Therefore, in addition to reasonable neatness, state any assumptions that you consider necessary to allow you to work the problem properly, and provide sufficient explanation so that the examiner can judge your reasoning. Assumptions should, of course, follow the logic and requirements of the problem.

Unfortunately, many electrical engineers find that the scope of the professional electrical engineering examination is somewhat broader than their own day-to-day activities. As a result it is desirable for most engineers to undertake a substantial technical review program before attempting the electrical engineering principles and practice examination. This book has been prepared to help engineers to organize their review effort.

2

BASIC CIRCUIT ANALYSIS

CIRCUITS 1.

Consider the following expression for a wave:

$f(t) = 10.0 \sin wt + 2.0 \cos (3wt + 90°)$

(1) Wt. 3 On the grid provided sketch the wave, f(t) to the scale given by locating ordinates for every 30° on the wt scale.

(Example) For the statement: The maximum value of the fundamental component of the wave, f(t), is:

a. - 0.0 c. - 10.0
b. - 2.0 d. - None of these

The designation, "c." for 10.0, should be checked.

(2) Wt. 2 The d-c component of the wave, f(t), is:

a. - 0.0 f. - 3.0
b. - 1.0 g. - 4.0
c. - 1.5 h. - 5.0
d. - 2.0 i. - 10.0
e. - 2.5 j. - None of these

(3) Wt. 3 The half-period average of the wave, f(t), is:

a. - 0.0 f. - 6.36
b. - 1.20 g. - 7.64
c. - 2.00 h. - 10.0
d. - 3.28 i. - 12.0
e. - 5.96 j. - None of these

(4) Wt. 2 The rms or effective value of the wave, f(t), is:

a. - 2.0 f. - 7.63
b. - 3.18 g. - 8.02
c. - 6.36 h. - 9.96
d. - 7.07 i. - 11.3
e. - 7.21 j. - None of these

(1) $f(t) = 10.0 \sin\omega t + 2.0 \cos(3\omega t + 90°)$
To determine values of the second term, set up a table:

ωt	$3\omega t$	$(3\omega t+90°)$	$\cos(3\omega t+90°)$
0	0	$\pi/2$	0
$\pi/6$	$\pi/2$	π	-1
$\pi/3$	π	$3\pi/2$	0
$\pi/2$	$3\pi/2$	2π	+1

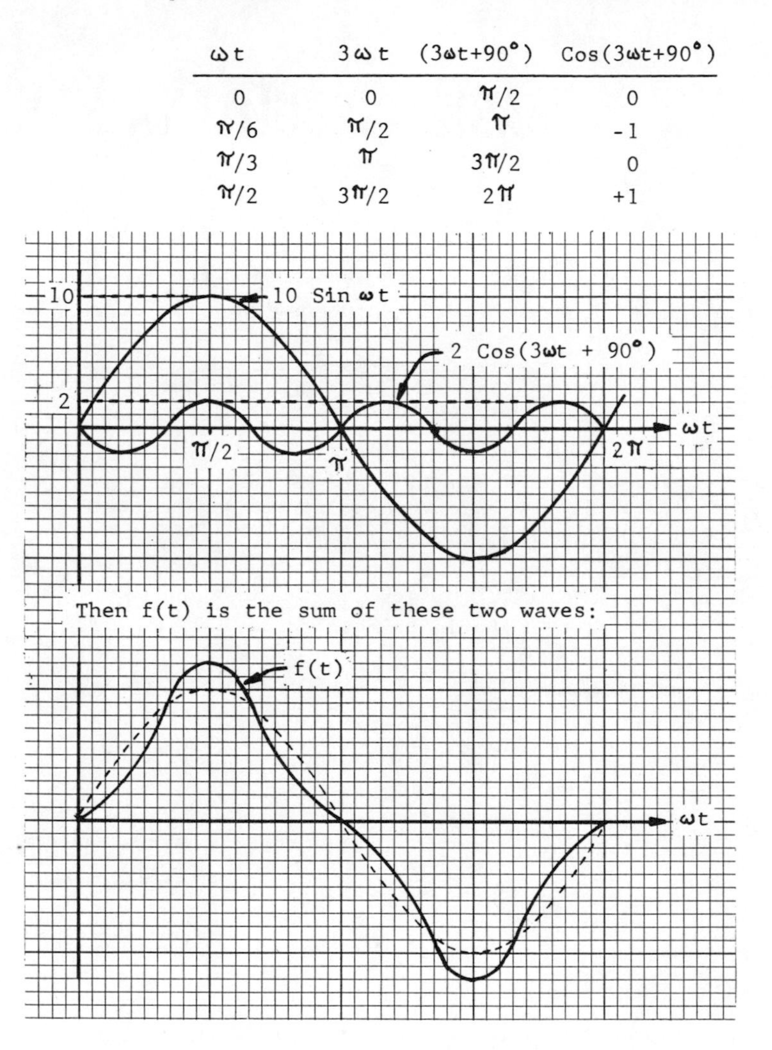

Then f(t) is the sum of these two waves:

(2) The d.c. component:

$$\text{d.c. component} = a_0 = \frac{1}{2\pi}\int_0^{2\pi} f(t)\, d(\omega t)$$

$$= \frac{1}{2\pi}\int_0^{2\pi} 10\text{Sin}\,\omega t\; d(\omega t) + \frac{1}{2\pi}\int_0^{2\pi} 2\text{Cos}(3\omega t + 90^\circ)\; d(\omega t)$$

$= 0+0 = 0.$ The answer is a. - 0.0

(3) The half-period average of the wave, f(t), is:

$$\text{d.c. half-period average} = a_{T/2} = \frac{1}{\pi}\int_0^{\pi} f(t)\, d(\omega t)$$

$$= \frac{1}{\pi}\left[\int_0^{\pi} 10\text{Sin}\,\omega t\; d(\omega t) + \frac{1}{3}\int_0^{\pi} 2\text{Cos}(3\omega t + 90^\circ)\; d(3\omega t)\right]$$

$$= \frac{1}{\pi}\left[+10 + 10 - 2/3 - 2/3\right] = \frac{1}{\pi}\left[20 - 4/3\right] = \frac{18.67}{\pi}$$

$= 5.95$ The answer is e. - 5.96

Note that there is one more negative than positive third harmonic loop per half-cycle.

(4) The rms or effective value of the wave, f(t), is:

$$\text{effective value of fundamental} = F_1 = \frac{10}{\sqrt{2}}$$

$$\text{effective value of 3rd harmonic} = F_3 = \frac{2}{\sqrt{2}}$$

$$F_{TOT(\text{effective})} = \sqrt{F_1^2 + F_3^2} = \sqrt{\frac{100}{2} + \frac{4}{2}} = 7.21$$

The answer is e. - 7.21

Consider again the expression for the wave in Problem I as follows:
$f(t) = 10.0 \sin wt + 2.0 \cos (3wt + 90°)$

(1) Wt. 2 The maximum or peak value of the wave, f(t), is:

a. - 2.0
b. - 4.0
c. - 6.0
d. - 8.0
e. - 10.0
f. - 11.5
g. - 12.0
h. - 14.0
i. - 16.0
j. - None of these

(2) Wt. 2 If the wave, f(t), represents a current in amperes that is flowing through a resistor having a resistance of 3.00 ohms, the power loss in watts is:

a. - 43
b. - 55
c. - 78
d. - 93
e. - 110
f. - 156
g. - 187
h. - 221
i. - 312
j. - None of these

(3) Wt. 3 If the wave, f(t), represents a current in amperes that is flowing through an inductor having an inductance of 0.1 henry, the rms or effective voltage in volts across the inductor is:

a. - 0.00
b. - 0.33
c. - 0.67
d. - 0.83
e. - 1.00
f. - 0.33w
g. - 0.67w
h. - 0.83w
i. - 1.00w
j. - None of these

(4) Wt. 3 If the wave, f(t), represents a current in amperes having an angular velocity of 6000 radians per second which current is passing through a capacitor having negligible losses, no initial charge, and a capacitance of 0.002 farads; the rms or effective voltage in volts across the capacitor is:

a. - 0.0
b. - 0.21
c. - 0.59
d. - 0.74
e. - 1.00
f. - 2.10
g. - 5.90
h. - 7.40
i. - 10.0
j. - None of these

(1) Note from the sketch the fundamental and the 3rd harmonic peaks occur in phase, thus $f(t)_{peak}$ is merely:

$$f(t)_{peak} = 10 + 2 = 12$$

The answer is g. - 12.0

(2) From Problem I, part (4):

$$I_{eff.} = \sqrt{I_1^2 + I_3^2} = 7.21 \text{ amperes}$$

then:

$$P = I_{eff}^2 R = (7.21)^2(3.00) = 156 \text{ watts}$$

The answer is f. - 156

(3) The effective voltage across the inductor:

$$E_{eff.} = \sqrt{E_1^2 + E_3^2}$$

where $E_1 = I_1 X_{L1} = \left(\frac{10}{\sqrt{2}}\right)\left(\omega L\right) = \frac{1.0}{\sqrt{2}}\omega$

and $E_3 = I_3 X_{L3} = \left(\frac{2}{\sqrt{2}}\right)\left(3\omega L\right) = \frac{0.6}{\sqrt{2}}\omega$

$$\therefore E_{eff.} = \sqrt{\frac{\omega^2}{2} + \frac{.36\omega^2}{2}} = \frac{\omega}{\sqrt{2}}\sqrt{1 + .36} = 0.83\,\omega$$

The answer is h. - 0.83 ω

(4) For $\omega = 6000$ radians/second:

$$E_1 = X_{C1} I_1 = \left(\frac{1}{\omega C}\right)\left(\frac{10}{\sqrt{2}}\right) = \frac{10}{(6\text{x}10^3)(2\text{x}10^{-3})\sqrt{2}} = \frac{10}{12\sqrt{2}}$$

$$E_3 = X_{C3} I_3 = \left(\frac{1}{3\omega C}\right)\left(\frac{2}{\sqrt{2}}\right) = \frac{2}{3\text{x}6\text{x}10^3\text{x}2\text{x}10^{-3}} = \frac{2}{36\sqrt{2}}$$

$$\therefore E_{eff.} = \sqrt{\left(\frac{10}{12\sqrt{2}}\right)^2 + \left(\frac{2}{36\sqrt{2}}\right)^2} = 0.59$$

The answer is c. - 0.59

CIRCUITS 3.

Subject: Circuits: Transients

A 2.0 μf capacitor is charged so that it has 100 volts across its terminals. The terminals are suddenly connected through a negligible resistance (actually by means of copper bars, 1/2" square and 10" long) to the terminals of a 4.0 μf capacitor having no initial charge.

(1) Wt. 2 What are the final steady state voltages across each capacitor?

(2) Wt. 1 What are the initial stored energies of each capacitor?

(3) Wt. 1 What are the final steady state stored energies of each capacitor?

A 100 ohm resistor is placed in series with the 4.0 μf capacitor so that the charging current will flow through the resistor. The same 2.0 μf capacitor charged to 100 volts is connected to the combination of the 4.0 μf capacitor and 100 ohm resistor in series with the 4.0 μf capacitor having no initial charge.

(4) Wt. 2 What is the time constant of the circuit?

(5) Wt. 1 What are the final steady state voltages across each capacitor?

(6) Wt. 1 What are the final steady state stored energies of each capacitor?

(7) Wt. 2 How do you explain or account for the results of the energies calculated in (3) and (6) above in the light of the different resistor power losses?

$Q = CE = 2 \times 10^{-6} \times 100 = 200 \times 10^{-6}$ coulomb

Capacitors are suddenly connected together, assuming no circuit resistance and the 4 μfd capacitor has $Q_2 = 0$.

The capacitors are in parallel across V so electrons will flow until equilibrium is reached.

The total charge $Q = 200 \times 10^{-6}$ coulomb remains in the system.

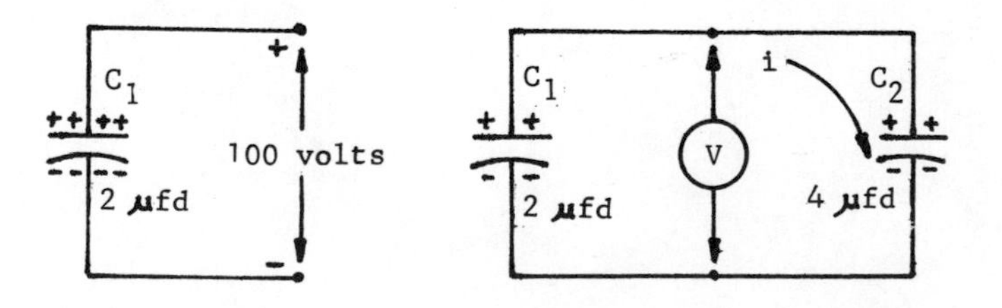

So $V = \frac{Q}{C_1 + C_2} = \frac{200 \times 10^{-6}}{(2 + 4)\times 10^{-6}} = 33\ 1/3$ Volts

(1) The voltage across each capacitor is 33 1/3 volts.

(2)

$$W = \int \frac{1}{c}\, dq = \frac{1}{2}\frac{Q^2}{C} \text{ and initially } 0$$

$Q_1 = 200 \times 10^{-6}$ coulomb so:

$$W_1 = \frac{1}{2}\frac{(200 \times 10^{-6})^2}{2 \times 10^{-6}} = \frac{(2 \times 10^{-4})^2}{4 \times 10^{-6}}$$

$$W_1 = \frac{1 \times 10^{-8}}{10^{-6}} = 1 \times 10^{-2} \text{ Joules}$$

energy initially in C_1

Q_2 was zero initially, so $W_2 = 0$

(3) Final energy

$Q_1 = 2 \times 10^{-6} \times 33\ 1/3 = 66\ 2/3 \times 10^{-6}$ coulomb

$$W_1 = \frac{1}{2}\frac{(66\ 2/3 \times 10^{-6})^2}{2 \times 10^{-6}} = \frac{4.45 \times 10^{-10}}{4 \times 10^{-6}}$$

$= 1.110 \times 10^{-3}$ Joules in C_1

$Q_2 = 4 \times 10^{-6} \times 33\ 1/3 = 133\ 1/3 \times 10^{-6}$

$$W_2 = \frac{1}{2}\frac{(133\ 1/3 \times 10^{-6})^2}{4 \times 10^{-6}} = \frac{1.76 \times 10^{-8}}{8 \times 10^{-6}}$$

$= 2.22 \times 10^{-3}$ Joules in C_2

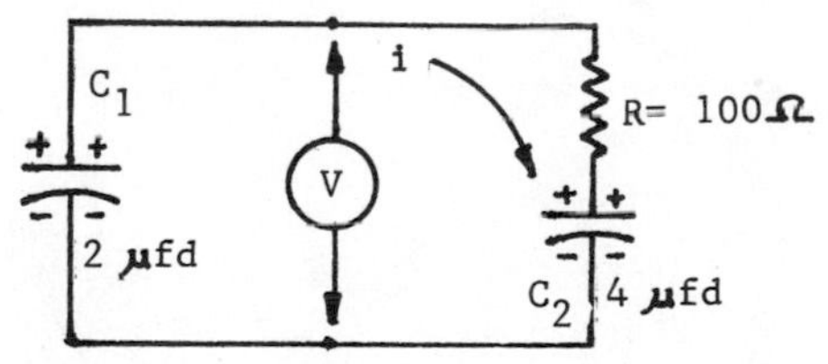

Time Constant $t = RC$

$$Ri + \frac{1}{C_1}\int i\,dt + \frac{1}{C_2}\int i\,dt = 0$$

i steady state = 0, so:
i transient is the solution of

$$R\frac{di}{dt} + \frac{1}{C_1}i + \frac{1}{C_2}i = 0 \qquad R\frac{di}{dt} + \left(\frac{C_1 + C_2}{C_1 C_2}\right)i = 0$$

$$\text{so } i = K\epsilon^{-\left(\frac{C_1 + C_2}{C_1 C_2 R}\right)t} \qquad \text{where } \frac{C_1 + C_2}{C_1 C_2 R} = \frac{2 + 4}{2 \times 4R} = \frac{3}{4R}$$

$$i = \frac{V_{C_1}}{R}\epsilon^{-\left(\frac{C_1 + C_2}{C_1 C_2 R}\right)t} \qquad \text{so } \frac{t}{RC} = \frac{t}{100 \times \frac{4}{3}} = 1$$

(4) Time Constant $t = RC = 100 \times 1.33 \times 10^{-6}$
$= 133 \times 10^{-6}$ Seconds

(5) $E_{C_1} = E_{C_2} = 33\ 1/3$ Volts, same as part (1).

(6) $W_1 = \frac{1}{2}\frac{(66\ 2/3 \times 10^{-6})^2}{2 \times 10^{-6}} = 1.11 \times 10^{-3}$ Joules

$W_2 = \frac{1}{2}\frac{(133\ 1/3 \times 10^{-6})^2}{4 \times 10^{-6}} = 2.22 \times 10^{-3}$ Joules

(7)
The circuit resistance determines the peak discharge or charge current, so there is I^2R loss in parts 1, 2, & 3 even if the resistance of the copper bar seems negligible. This accounts for the loss in energy.

CIRCUITS 4.

The circuit shown below is in a steady state.

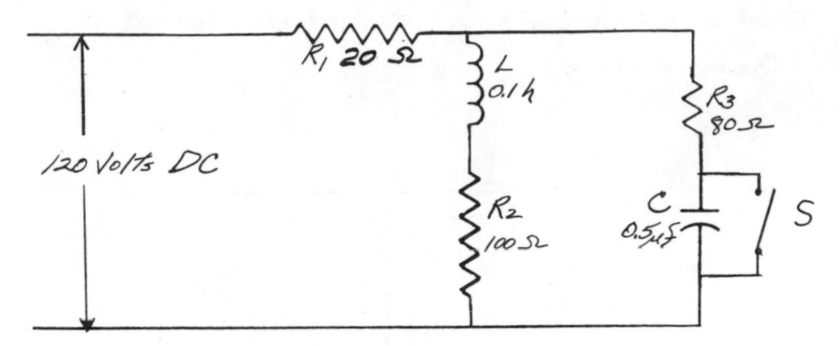

REQUIRED:

Wt.		
1	(a)	What current is drawn from the power source in this initial steady state condition?
9	(b)	What is the analytical expression for the current drawn from the power source after closing switch S?

a) With the switch open, and with the statement that the circuit is in steady state (to a dc source), one may make the assumption that the current through the inductor is no longer changing; the voltage across this element will then be zero. The voltage across the capacitor has built up to its steady state value and therefore no current will be flowing in this branch:

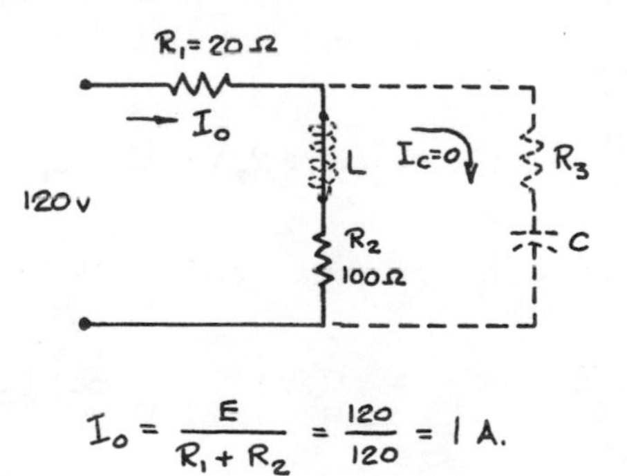

$$I_o = \frac{E}{R_1 + R_2} = \frac{120}{120} = 1 \text{ A.}$$

b) After the switch is closed, the equivalent circuit will be as follows (with an initial inductor current as found in part a):

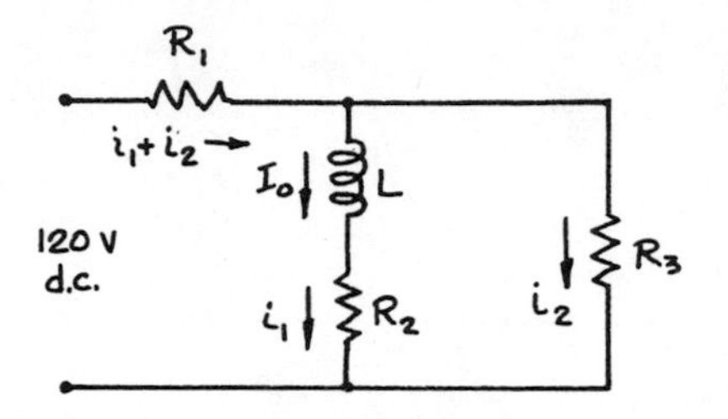

Loop (differential) equations:

$$1) \quad E = R_1(i_1 + i_2) + L\frac{di_1}{dt} + R_2 i_1$$

$$2) \quad E = R_1(i_1 + i_2) + R_3 i_2$$

with an initial condition of $i_1(0) = I_0$

Using Laplace transforms to solve:

$$1) \quad \frac{E}{s} = R_1(I_1 + I_2) + L[sI_1 - i_1(0)] + R_2 I_1$$

$$2) \quad \frac{E}{s} = R_1(I_1 + I_2) + R_3 I_2$$

Rearranging terms:

$$1) \quad \frac{E}{s} + L\,i_1(0) = [(R_1 + R_2) + Ls]\,I_1 + R_1 I_2$$

$$\frac{120}{s} + 0.1 = (120 + 0.1s)\,I_1 + 20\,I_2$$

$$2) \quad \frac{120}{s} = (20)\,I_1 + (100)\,I_2$$

Solving for I_1 and I_2:

$$I_1 = \frac{\begin{vmatrix} \left(\frac{120}{S} + 0.1\right) & 20 \\ \left(\frac{120}{S}\right) & 100 \end{vmatrix}}{\begin{vmatrix} (120 + 0.1S) & 20 \\ 20 & 100 \end{vmatrix}} = \frac{\left(\frac{120}{S} + 0.1\right)100 - \left(\frac{120}{S}\right)20}{(120 + 0.1S)\,100 - 20^2}$$

$$= \frac{96 + 0.1S}{S(116 + 0.1S)}$$

$$I_2 = \frac{\begin{vmatrix} (120 + 0.1S) & \left(\frac{120}{S} + 0.1\right) \\ 20 & \left(\frac{120}{S}\right) \end{vmatrix}}{(120 + 0.1S)100 - 20^2} = \frac{120 + 0.1S}{S(116 + 0.1S)}$$

But $I_{source} = I_1 + I_2$

$$\therefore\ I_s = \frac{96 + 0.1S + 120 + 0.1S}{S(116 + 0.1S)} = \frac{216 + 0.2S}{S(116 + 0.1S)}$$

$$= \frac{216}{116}\left[\frac{1 + \frac{0.2}{216}S}{S\left(1 + \frac{0.1}{116}S\right)}\right] = 1.86\left[\frac{1 + \tau_1 S}{S(1 + \tau_2 S)}\right]$$

where $\tau_1 = 0.000927$

$\tau_2 = 0.0008625$

$$\therefore\ i_s(t) = \mathcal{L}^{-1}[I_s] = 1.86\left[1 - (1 - \tau_1/\tau_2)e^{-t/\tau_2}\right]$$

$$= 1.86\,(1 + 0.072\,e^{-t/\tau_2})$$

$.133\,e^{-t/\tau_2}$

CIRCUITS 5.

Refer to the circuit below. If the inductance L is 10.0 microhenries and the frequency of the applied voltage is 9.55 megacycles, determine the value of the reactance of C so that the circuit will be series resonant. What is the impedance looking into the circuit under these conditions?

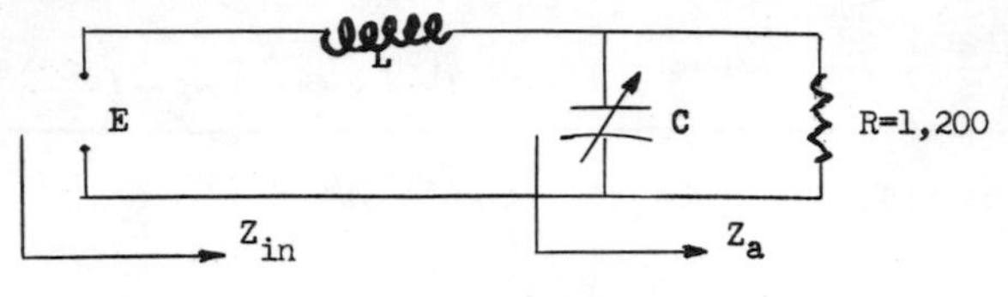

(a) Solve circuit by the admittance method:

$$Y_a = \frac{1}{Z_a} = \frac{1}{1,200} + {}_jYca = 0.833 \times 10^{-3} + {}_jYca$$

$$Z_a = \frac{1}{0.833 \times 10^{-3} + {}_jYca} = \frac{0.833 \times 10^{-3}}{D} - \frac{{}_jYca}{D} = R_a - {}_jXca$$

Where denominator = D after rationalizing the fraction with the conjugate ; $D = 0.693 \times 10^{-6} + Y^2_{ca}$,

Using: $(a-b)(a+b) = a^2 - b^2$.

New equivalent circuit:

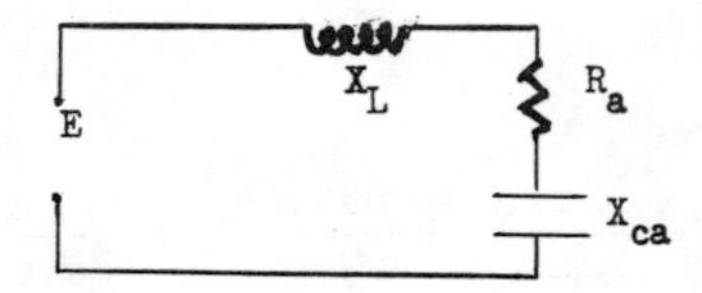

At Resonance: $X_{ca} = X_L$

$$X_L = 2\pi fL = (6.28)(9.55 \times 10^6)(10 \times 10^{-6}) = 599.7 \text{ ohms}$$

Thus $\frac{Yca}{D} = 599.7 = X_{ca}$ based on the principles of resonance.

If $\frac{Yca}{D} = 599.7$, then

$$Yca = 599.7(0.693 \times 10^{-6} + Y^2_{ca})$$

$$599.7Y^2_{ca} - Y_{ca} + 4.156 \times 10^{-4} = 0$$

Solving the seoond degree equation by using the standard formula we obtain:

$$Y_{ca} = \frac{1 \pm \sqrt{1 - 4\ (4.156 \times 10^{-4})\ (599.7)}}{2 \times 599.7}$$

$$= \frac{1 \pm \sqrt{1 - 0.9979}}{2 \times 599.7} = \frac{1}{1{,}199.4} = 0.833 \times 10^{-3}$$

Where the radical was approximated to be zero

Therefore the reactance requested is:

$$X_{ca} = \frac{1}{Y_{ca}} = 1{,}199.4 \text{ ohms}$$ ANS.

(b) $Z_{in} = R_a$, as in resonance the only effective part of the impedance is the real part of the complex expression

$$Z_{in} = R_a = \frac{0.833 \times 10^{-3}}{D} = \frac{0.833 \times 10^{-3}}{0.693 \times 10^{-6} + Y_{ca}^2} =$$

$$\frac{0.833 \times 10^{-3}}{0.693 \times 10^{-6} + (0.833 \times 10^{-3})^2}$$

$$Z_{in} = 0.6 \times 10^3 = 600 \text{ ohms}$$ ANS.

CIRCUITS 6.

The switch was closed sufficiently long ago that the current i (through the source) has reached steady state. The switch is then opened at time $t = t_o$.

REQUIRED

Wt

1 (a) Find the current i just before switch is opened; that is, at $t = t_{o}-$

3 (b) Find the current i just after the switch is opened; that is, at $t = t_{o}+$

6 (c) Find the current i as a function of time after the switch is opened; that is, find $i\ (t - t_o)$.

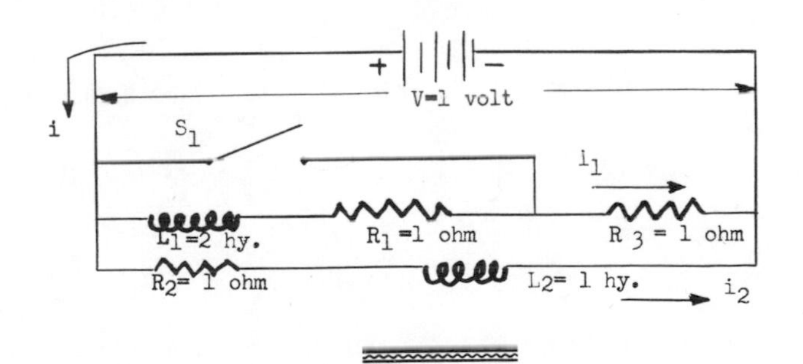

(a) At $t = t_o-$ only R_3 and R_2 determine the magnitude of i

$$R_{total} = \frac{1 \times 1}{1 + 1} = 0.5 \text{ ohm}$$

$$i = \frac{E}{R_{total}} = \frac{1}{0.5} = 2 \text{ amp} \qquad \text{ANS.}$$

(b) At $t = t_o+$, since the current in an inductance will not change instantaneously, the current in L_1 will remain at 0. Only R_2 will determine the magnitude of i

$$i = \frac{E}{R_2} = \frac{1}{1} = 1 \text{ amp} \qquad \text{ANS.}$$

(c) At $(t - t_o)$ the R_2 and L_2 branch of the circuit is in steady state and is not considered. Therefore for the L_1, R_1 and R_3 branch:

$$i = i_{transient} + i_{R_2}$$

$$i_{transient} = \frac{E}{R} \left(1 - e^{-\frac{Rt}{L_1}} \right)$$

$$i_{R_2} = 1 \text{ amp (See above (b))}$$

$$R = R_1 + R_3 = 1 + 1 = 2 \text{ ohms}$$

Therefore

$$i = \frac{1}{2} - \frac{1}{2} e^{-t} + 1 = 1.5 - 0.5 e^{-t} \qquad \text{ANS.}$$

Proof: at t = o .... $e^{-t} = 1.0$ and $i = 1.5 - 0.5 \times 1.0 =$ 1 amp, q.e.d.

<u>NOTE</u>: Part (c) can be also worked by using the classic solutions of:

$$L_1 \frac{di_1}{dt} + (R_1 + R_3)\, i_1 = 1$$

$$L_2 \frac{di_2}{dt} + R_2\, i_2 = 1$$

$i = i_1 + i_2 = 1.5 - 0.5e^{-t}$ by applying the proper boundary conditions.

CIRCUITS 7.

An alternating current voltmeter consists of a series connection of an ideal half-wave diode and a D'Arsonval meter. The meter is calibrated to read the rms value of an applied voltage. When the waveform sketched below is applied, the meter reads 80 volts. What is the peak value of the applied waveform?

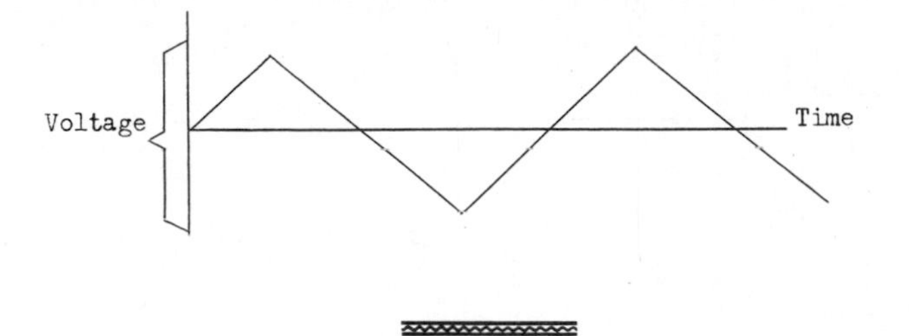

The voltage as seen by the D'Arsonval meter (assuming a lossless diode rectifier) from a pure sine wave: $e = E_{max} \sin \omega t$

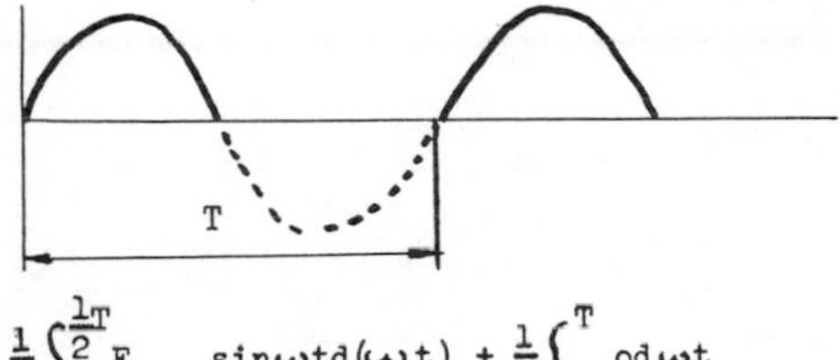

Then: $E_{avg} = \frac{1}{T}\int_{0}^{\frac{1}{2}T} E_{max} \sin\omega t\, d(\omega t) + \frac{1}{T}\int_{\frac{1}{2}T}^{T} 0\, d\omega t$

Actual voltage read would be E_{max}/π but meter is calibrated to read $E_{max}/\sqrt{2}$ (for an rms value).

For the wave shape given:

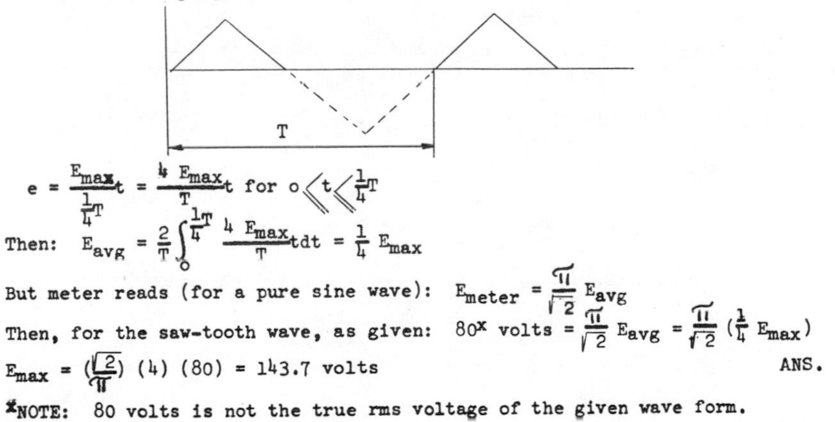

$e = \frac{E_{max}}{\frac{1}{4}T} t = \frac{4\, E_{max}}{T} t$ for $0 \leq t \leq \frac{1}{4}T$

Then: $E_{avg} = \frac{2}{T}\int_{0}^{\frac{1}{4}T} \frac{4\, E_{max}}{T} t\, dt = \frac{1}{4} E_{max}$

But meter reads (for a pure sine wave): $E_{meter} = \frac{\pi}{\sqrt{2}} E_{avg}$

Then, for the saw-tooth wave, as given: 80^{x} volts $= \frac{\pi}{\sqrt{2}} E_{avg} = \frac{\pi}{\sqrt{2}} (\frac{1}{4} E_{max})$

$E_{max} = (\frac{\sqrt{2}}{\pi})\ (4)\ (80) = 143.7$ volts ANS.

*NOTE: 80 volts is not the true rms voltage of the given wave form.

CIRCUITS 8.

REQUIRED:

Wt.

5 (a) In the bridge circuit shown below, find the current through the meter as a function of R, the deviation of the unknown resistor from 1000 ohms.

3 (b) Find the points at which the meter should be marked if the bridge is to be used to reject resistors which exceed 10% tolerance ($|R| \leq 100$).

2 (c) Discuss the effect of battery aging on the tolerance, assuming that the marks are not readjusted.

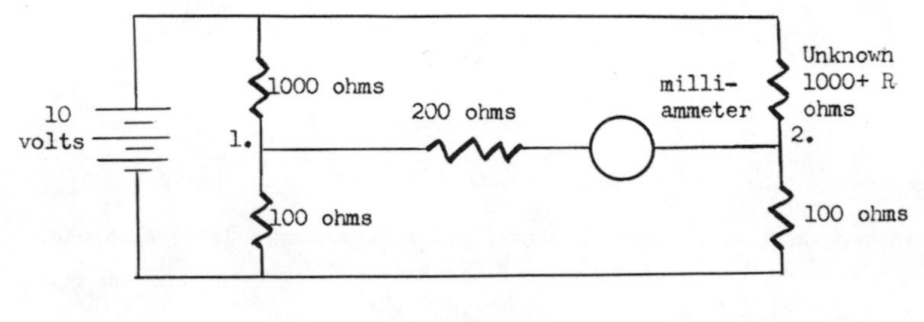

(a) Based on the principle of voltage dividers (for finding Thevenin's open circuit voltage):

$$V_1 = \frac{100 \times 10}{1,100} \text{ and } V_2 = \frac{100 \times 10}{100 + P}$$

Where P = 1000 + R

$$V_{12} = V_1 - V_2 = 10^3 \left(\frac{1}{1,100} - \frac{1}{100 + P}\right) = 10^3 \left[\frac{100 + P - 1,100}{1,100\ (100 + P)}\right]$$

$$= 10^3 \left[\frac{P - 1000}{1,100\ (P + 100)}\right] = \frac{P - 1000}{1.1\ (P + 100)}$$

Based on Thevenin theorem:

$$Z_{12} = \frac{100 \times 1,000}{1,100} + \frac{100\ P}{100 + P} = \frac{10 \times 10^6 + 10^5 P + 11 \times 10^4 P}{1,100\ (100 + P)}$$

$$= \frac{10^3\ (10 + 0.21P)}{110 + 1.1P}$$

$$Z_{12} + 200 = \frac{10,000 + 210P + 22,000 + 220P}{110 + 1.1P}$$

$$= \frac{430P + 32,000}{1.1P + 110} = \frac{430P + 3,200}{1.1\ (P + 100)}$$

$$I_a = \frac{V_{12}}{Z_{12} + 200} = \frac{P - 1000}{1.1\ (P + 100)} \times \frac{1.1\ (P + 100)}{430P + 32,000} = \frac{P - 1000}{430P + 32,000}$$

Substituting 1,000 + R = P we obtain:

$$I_a = \frac{R}{430\ R + 462{,}000}$$ ANS.

(b) If R= +100

Mark one meter point: $I_a = \frac{100}{43{,}000 + 462{,}000} = \frac{100}{505{,}000}$

$= 0.198 \times 10^{-3}$ Amps ANS.

If R = -100

Mark second meter point: $I_a = \frac{-100}{-43{,}000R + 462{,}000} = \frac{-100}{41{,}900}$

$= -0.238 \times 10^{-3}$ Amps ANS.

(c) As the battery ages, its internal resistance increases, reducing all voltages including the voltage differences between points 1 and 2 across which the current is being measured. The range of acceptable current readings would be therefore smaller. If the marks are not reduced accordingly, some resistors would be accepted that should be rejected. ANS.

3
FIELDS, TRANSMISSION LINES, & MAGNETIC CIRCUITS

PROBLEM 1.

A transducer whose impedance is 600 ohms resistive generates 0.1 volt rms at a frequency of 100 Hz. It is desired to transmit this signal over a telephone cable whose length is 2.5 miles. The characteristic impedance of the line, Z_o, is 600 ohms, and its loss is 0.40 neper per mile. The phase shift is 0.1255 radians per mile. All parameters are measured at 1000 Hz. The input to the line is to be amplified by amplifier A so as to feed the line at a level of 0 dbm. The output of the line is to be amplified by amplifier B so as to provide 2.0 watts to a recorder. Amplifiers A and B each have an input impedance of 600 ohms.

REQUIRED:

(a) What is the gain of amplifier A in decibels?

(b) What is the gain of amplifier B in decibels?

(c) What is the delay in microseconds between the input and output of the line?

(d) What would be the least expensive way to reduce the 1000 Hz loss in the telephone line?

(e) What effect, if any, would this measure in (d) above have on the line's delay?

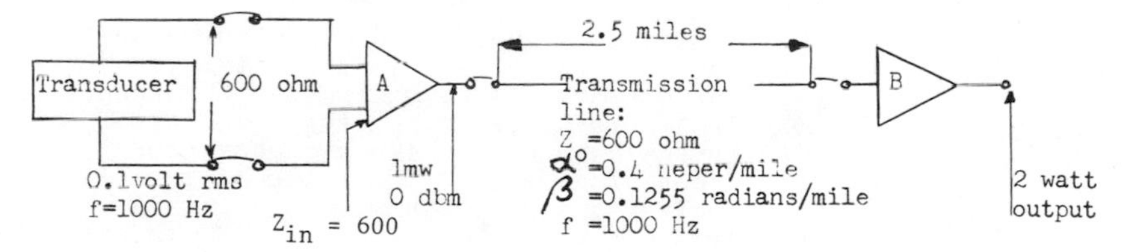

(a) P_{in} level $= \frac{E^2}{R} = \frac{(0.1)^2}{600} \times 10^3 = \frac{0.01}{600} \times 10^3 = 0.0167$ mw

P_{out} level is given: 0dbm = 1_{mw}

$$\text{db Gain}_A = 10 \log_{10} \frac{P_{out}}{P_{in}} = 10 \log_{10} \frac{1}{0.0167} = 10 \log_{10} 60$$

$$= 10 \times 1.778 = 17.78\text{db}$$ ANS.

(b) Loss in cable: 0.4 nepers/mile x 2.5 miles = 1 neper

1 neper x 8.686 db/nepers = 8.686 db

P_{in} level = 0 - 8.686 = -8.686 dbm

P_{out} level = $10 \log \frac{2}{1 \times 10^{-3}} = 10 \log 2 \times 10^3 = 10 \times 3.3 = 33$ dbm

db $Gain_B$ = 33.0 - (-8.686) = 41.686db ANS.

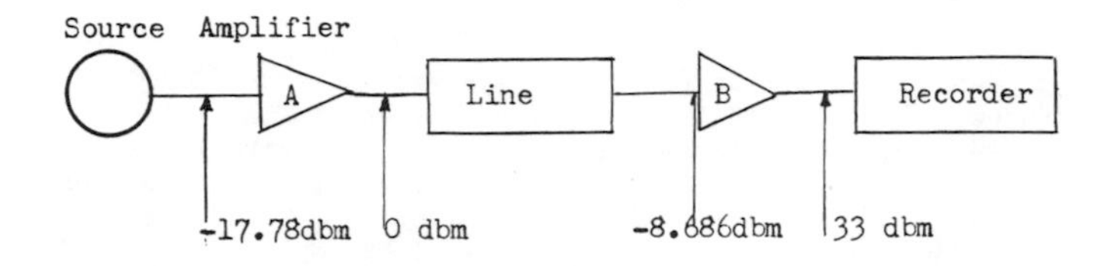

(c) Delay = $\frac{0.1255 \text{ rad/mi} \times 2.5\text{mi} \times 10^6 \frac{\text{microsec}}{\text{sec}}}{2\pi \times 1000 \text{ H}_z}$ = 50 microseconds ANS.

(d) The loss at 1000 Hz could be reduced the least expensively by loading (series) the telephone line; i.e., inserting inductance coils at regularly spaced intervals.

(e) The delay of the transmission line will increase when load coils are added, since we increased the phase shift.

PROBLEM 2.

Subject: Magnetic circuits: Lifting magnet

A circular lifting magnet for a crane is to be designed so that with a flux density of 30,000 lines per square inch in each air gap, the length of each air gap is 0.5 inch. Leakage and saturation effects are such that the magnetomotive force for the air gaps is 0.85 of the magnetomotive force for the complete magnetic circuit. The mean length per turn of winding is 24 inches and the magnet is to operate with an applied terminal voltage of 70 volts with the winding at a temperature of 60°C.

(1) Wt. 2 What is the pull or tractive force in pounds per square inch for the magnet?

(2) Wt. 8 What size wire should be used for the winding?

Assume a d-c magnet.

Force in dynes $= \frac{B^2 A}{8\pi}$ Maxwell's Eqn

Force in lbs/sq. in. $= \frac{B^2}{72\times10^6} = \frac{(30,000)^2}{72\times10^6}$

(1) Force = 12.5 lbs/sq. in. pull

(2)

$B = 30,000$ Lines/sq. in., $L = 0.5$ in. air gap

$mmf_{air\ gap} = 0.85$ total NI

L_{mean} of coil = 24 inches

V = 70 volts on coil

$H_{air\ gap} = 0.313$ B NI per inch for air gap

$NI_{for\ air\ gap} = 0.313\times30,000\times0.5\times2$ air gaps

= 9,400 Ampere turns for air gap

$NI_{total} = \frac{9,400}{0.85} = 11,050$ Ampere turns for air gap & magnetic ckt.

Since the coil dimensions are not given, assume 400 square inches of radiating surface and allow 0.7 watts per square inch dissipation at 60°C.

Watts = 400 x 0.7 = 280 watts dissipated

and $I = \frac{280 \text{ watts}}{70 \text{ volts}} = 4$ Amperes coil current

$N = \frac{NI}{I} = \frac{11050}{4} = 2763$ Turns on coil

Wire Length $= \frac{2763 \times 24 \text{ inches}}{12 \text{ inches/foot}} = 5526$ feet

$R_{wire} = \frac{70 \text{ volts}}{4 \text{ amperes}} = 17.5$ ohms

$R_{wire} = \frac{\rho \text{ length}}{\text{area}}$ where $\rho = 12$ at 60°C.

$A_{cir.\ mills} = \frac{12 \times 5526}{17.5} = 3880$ cir. mills

No. 14 AWG Magnet wire has 4,107 cir. mills and 2.525 ohms/1000'.

No. 15 AWG Magnet wire has 3,257 cir. mills and 3.184 ohms/1000'.

Choose No. 14 AWG Magnet wire.

PROBLEM 3.

Subject: Fields: Poynting vector

(1) Wt. 2 What is the Poynting vector? To what is it equal in terms of electric and magnetic quantities? Explain symbols or terms used carefully.

(2) Wt. 4 Consider the negative conductor of a two-wire, d-c, transmission line. Making free use of sketches, explain in words or demonstrate analytically how the Joulean energy loss is supplied by the Poynting vector. Indicate Poynting vector directions internal and external to the conductor.

(3) Wt. 4 Consider a 3-phase, 2-pole, 60-cps, squirrel cage induction motor the rotation of which is clockwise when viewed from the drive shaft end. Assume the motor is operating at full load and explain in words or demonstrate analytically how the energy made available in the stator circuit crosses or flows through the air gap of the machine by use of the Poynting vector. Free use of sketches should be made.

---

(1) Poynting's Law states that transfer of energy can be expressed as the product of the values of magnetic field and the component of the electric field perpendicular to the magnetic field. The energy flow is in a direction perpendicular to both fields at any point.

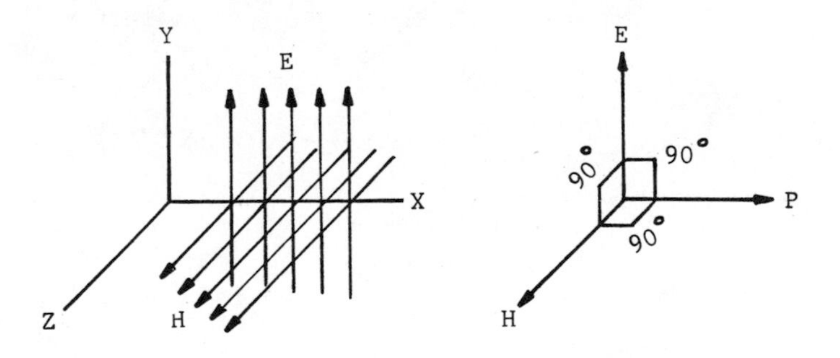

E is electric field intensity in volts per meter.

H is magnetic field in amperes per meter.

P is poynting vector in watts per square meter.

Energy flow is perpendicular to the E & H planes, i.e., the Y & Z planes, and is in watts per square meter. $\overline{P} = \overline{E} \times \overline{H}$ in vector form or $P = EH\sin\Theta$ where Θ is the angle between E & H.

(2) Consider an imaginary cylindrical surface about a conductor with a field "H" (due to the current) as shown and two components of an electric field "E", one component (up) representing the IR drop in the conductor, the other component represents the electric field between the lines. The poynting vector will then have a

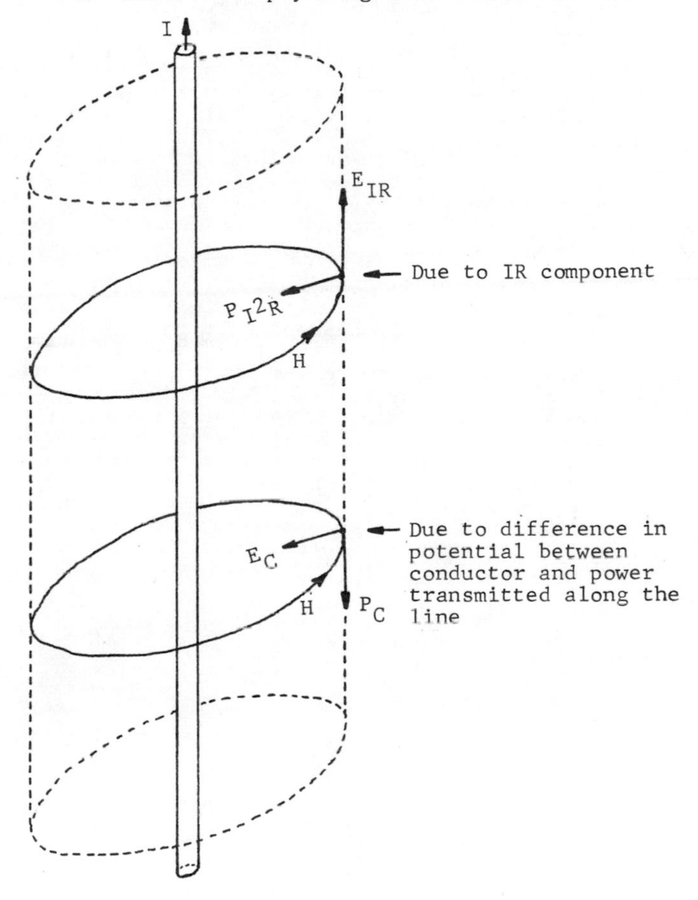

direction dependent on the electric field components, one component in towards the center of the conductor (representing the copper loss of the conductor), and the other in the direction of the conductor (representing the power transmitted to the load).

The poynting vector "$\overline{P}$" is then the vector sum of $\overline{P}_{I^2R} + \overline{P}_C = \overline{P}$.

(3) The 3Ø, 2 pole, 60 cycle squirrel-cage induction motor has 3Ø energy applied to the stator which produces a 2 pole 3 phase rotating field. The resultant flux vector of the 3Ø field travels clockwise around the stator at 3600 rpm. The squirrel-cage rotor in this field will rotate at almost synchronous speed at no load. The stator currents will produce an "H" field. The "E" field will be in the same plane as the "H" field. Perpendicular to the resultant "E" and "H" fields will be the resultant poynting vector "P". The instantaneous poynting vector has two components, the energy component and the I^2R loss component explained in part (2) of this problem.

The rotor delivering full load will have slip, the amount of slip depending on the N.E.M.A. class of motor. As a result of the slip, voltage will be induced in the rotor proportional to BLV. Rotor current will flow but in opposition to stator current as a result of the induced rotor voltage. If one of the rotor bars (conductors) is treated instantaneously, the poynting diagram will be exactly like the diagram in part (2) of this problem, on one side of the rotor. On the opposite side of the rotor the poynting vector would have an opposite direction.

PROBLEM 4.

A section of nonloaded telephone cable, located in a rural area, has the following characteristics per loop mile at a frequency of 1000 hertz:

Series resistance, 85.8 ohms
Inductance, 1.00 millihenry
Capacitance, 0.062 microfarad
Shunt conductance, 1.50 micromho

The cable consists of 400 pairs of No. 19 AWG, is shielded, and is 3.06" o.d.

REQUIRED:

Compute the following parameters for this cable at 1000 hertz:

Wt.		
3	(a)	characteristic impedance
3	(b)	attenuation in decibels per mile
2	(c)	phase shift per mile
2	(d)	velocity of propagation

Note: For full credit, you must show all of your work.
Answers taken directly from reference books are not acceptable.

(a)

The characteristic impedance of a line, or Z_c is a complex expression composed of:

z = series impedance per unit length, per phase = $R + j \times 2\pi f L$
y = shunt admittance per unit length, per phase to neutral = $G + j2\pi f c$

Converting to henry, farad and mho (multiply by 10^{-3} or 10^{-6})

$$z = 85.8 + j\ 2\pi \times 1000 \times 10^{-3} = 85.8 + j\ 6.28 = 85.9 \angle 4^\circ \text{ ohm / mile}$$

$$y = (1.5 + j\ 2\pi \times 1000 \times 0.062) \times 10^{-6} = (1.5 + j\ 389) \times 10^{-6} =$$

$$= 389 \times 10^{-6} \angle 89.9^\circ \text{ mho / mile}$$

$$Z_c = \sqrt{\frac{z}{y}} = \sqrt{\frac{85.9 \angle 4^\circ}{389 \times 10^{-6} \angle 89.9^\circ}} = \sqrt{0.221 \times 10^6} \angle \frac{4^\circ - 89.9^\circ}{2} =$$

$$= 4.7 \times 10^2 = \angle -43^\circ = 470 \angle -43^\circ \text{ ohms ANS}$$

(b)

Let us denote a complex quantity γ as the propagation constant and l the line length in miles:

$$\gamma = \sqrt{z\ l y l} = \sqrt{zyl^2} = l\sqrt{zy}$$

The real part of the propagation constant γ is called "attenuation constant" (β) and is measured in nepers per unit length; the quadrature part of (γ) is called the "phase constant" (β) and is measured in radians per unit length; thus,

$$\gamma l = l \times \sqrt{yz} = 1.0 \times \sqrt{85.9 \times 389 \times 10^{-6}} \angle \frac{4^\circ + 89.9^\circ}{2}$$

$$= 0.182 \angle 47^\circ = 0.124 + j\,0.133 \text{ (radians)} = \alpha + j\beta$$

Converting nepers into decibels, we obtain:

α = 0.124 nepers / mile = 0.124 x 8.686 = 1.08 decibels/mile ANS.

(c) Converting radians into degrees: 180° = 3.1416 radians, then 0.133 radians = 7.6°, thus phase shift or β = 7.6°/mile ANS.

(d) A wavelength is the distance along a line between two points of a wave which differ in phase by 360° or 2π radians. If λ is the phase shift in radians per mile, the wavelength in miles is

$$\lambda = \frac{2\pi}{\beta} = \frac{6.28}{0.133} = 47.2 \text{ miles}$$

Velocity of propagation is the product of the wavelength in miles and the frequency in cycles per second, or

velocity = λf = 47.2 x 1000 = 47,200 miles / second ANS.

Reference: Elements of Power System Analysis, by William D. Stevenson, Jr., McGraw Hill, p. 100-106.

PROBLEM 5.

Shown below is a diagram of a coaxial transmission line in which the circuit elements have the values indicated, and E_g is the open circuit generator voltage, Z_o is the characteristic impedance of all transmission lines, and L_2 is greater than one wavelength.

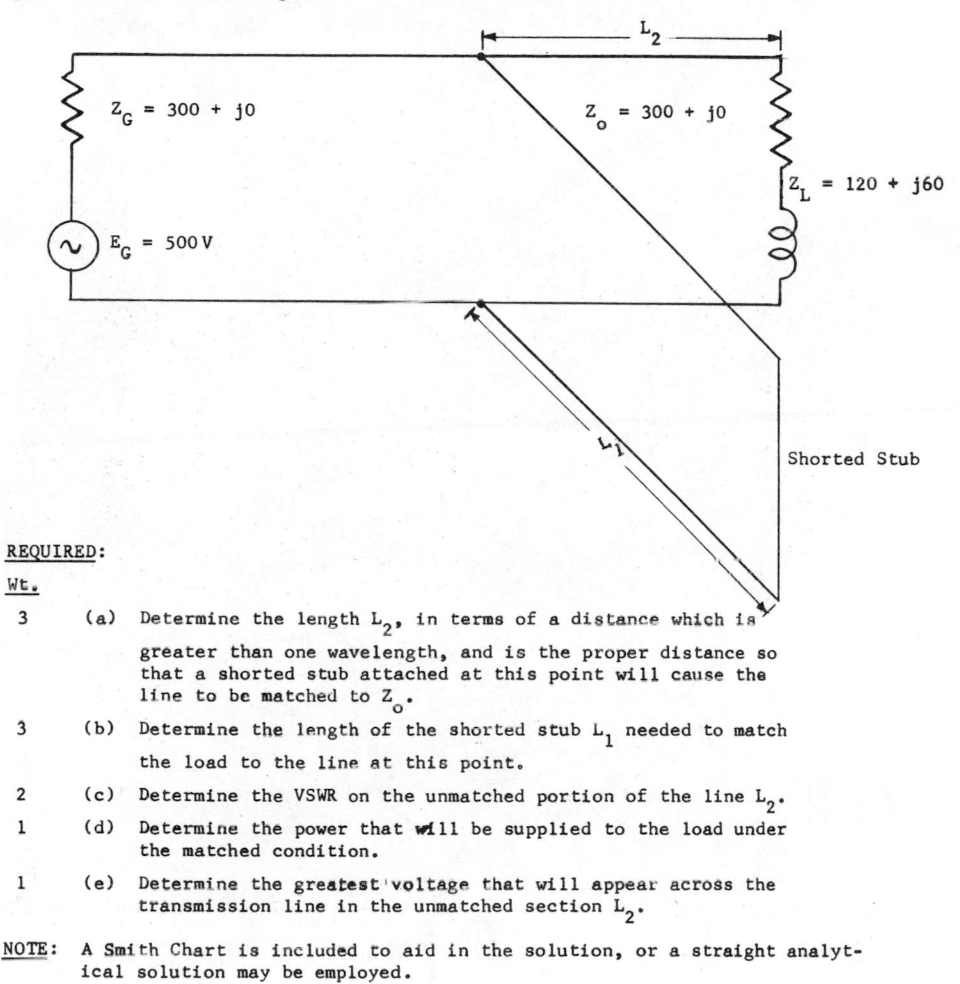

REQUIRED:

Wt.		
3	(a)	Determine the length L_2, in terms of a distance which is greater than one wavelength, and is the proper distance so that a shorted stub attached at this point will cause the line to be matched to Z_o.
3	(b)	Determine the length of the shorted stub L_1 needed to match the load to the line at this point.
2	(c)	Determine the VSWR on the unmatched portion of the line L_2.
1	(d)	Determine the power that will be supplied to the load under the matched condition.
1	(e)	Determine the greatest voltage that will appear across the transmission line in the unmatched section L_2.

NOTE: A Smith Chart is included to aid in the solution, or a straight analytical solution may be employed.

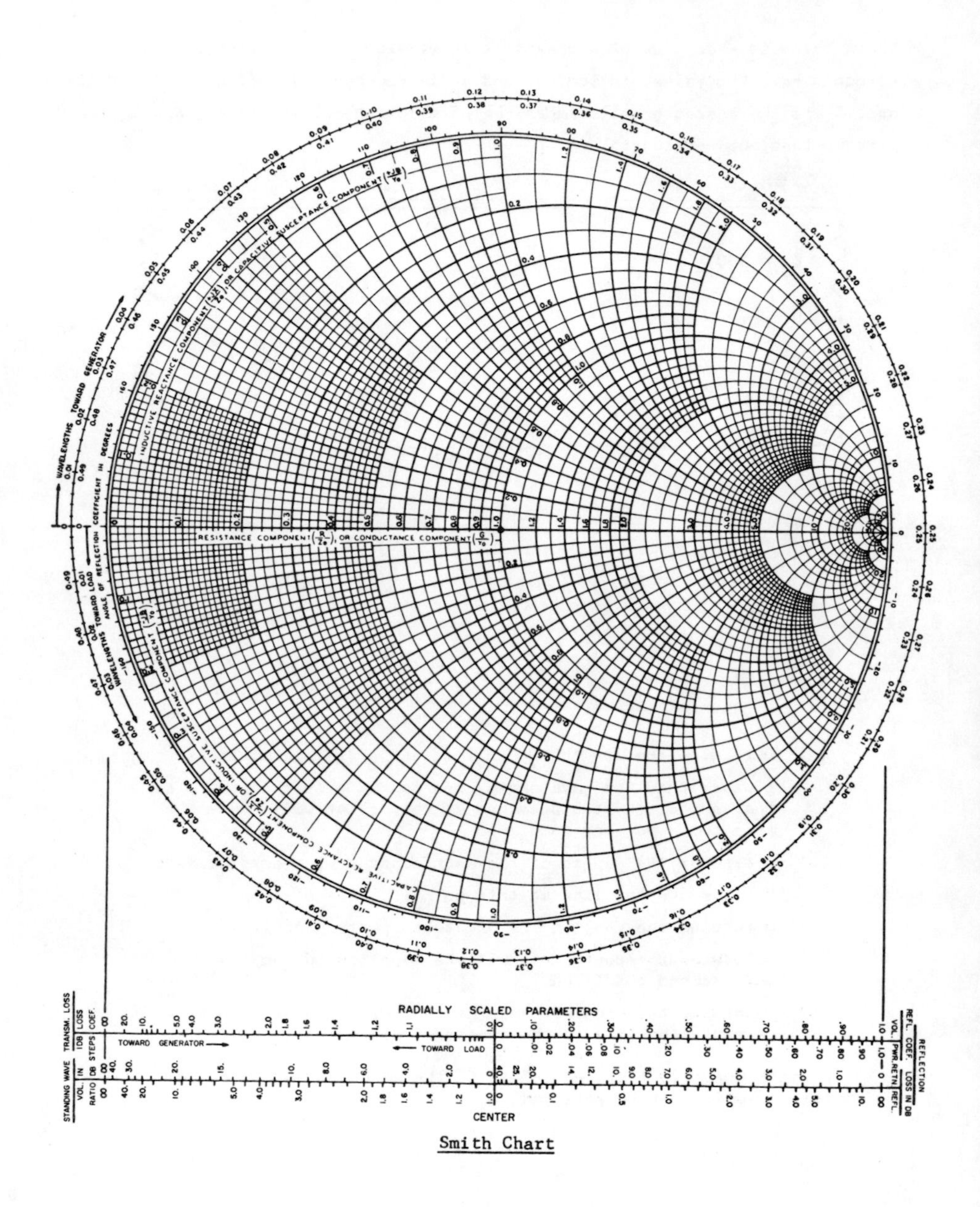
RESISTANCE COMPONENT, OR CONDUCTANCE COMPONENT
INDUCTIVE REACTANCE COMPONENT, OR CAPACITIVE SUSCEPTANCE COMPONENT
CAPACITIVE REACTANCE COMPONENT, OR INDUCTIVE SUSCEPTANCE COMPONENT
WAVELENGTHS TOWARD GENERATOR
WAVELENGTHS TOWARD LOAD
ANGLE OF REFLECTION COEFFICIENT IN DEGREES
RADIALLY SCALED PARAMETERS
TOWARD GENERATOR
TOWARD LOAD
STANDING WAVE
VOL. RATIO
IN DB
TRANSM. LOSS
1DB STEPS
LOSS COEF.
REFLECTION
REFL. COEF.
VOL.
PWR.
RETN LOSS IN DB
REFL. LOSS IN DB
CENTER

Smith Chart

As a first step, the Y_γ / Y_o ratio has to be determined:

$$Y_\gamma = \frac{1}{Z_\gamma} = \frac{1}{120 + j60} = \frac{1}{120 + j60} \times \frac{120 - j60}{120 - j60} = 0.0067 - j0.0033$$

$$Y_o = \frac{1}{Z_o} = \frac{1}{300} = 0.0033$$

$$\frac{Y}{Y_o} = \frac{0.0067 - j0.0033}{0.0033} = 2 - j1$$

This ratio is point A on the Smith Chart.

(a) To determine the length L_2 the stub is placed at the point where $G = 1/Z_o$ or at $G = 1$ which is point B on the Smith Chart. However, L_2 must be larger than one wavelength, so that location is 360° further away from load than indicated.

Initial location:	-26.5°
Final location:	-61.8°
Difference	35.3°

$$\frac{35.30°}{2} = 17.65°$$

The proper distance of the stub: $17.65° + 360° = 377.65°$ ANS.

(b) The length of the shorted stub L_1 is such that reactance of type opposite to line is required.

On Smith Chart start at U_1 ($Y_{\gamma\,stub}/Y_o = \infty$), move CCW to U_2 where the imaginary component of $G + jB$ is opposite that of stub location but equal in magnitude.

Initial location:	0°
Final location:	270°
Difference	270°

The length of the stub: $\frac{270°}{2} = 135°$ ANS.

(c) To determine the VSWR on the unmatched portion of the line L_2, from point C on Smith Chart, we obtain

$N = 2.6$ ANS.

This value can also be obtained as follows:

$$\rho = \frac{Z_L - Z_o}{Z_L + Z_o} = \frac{120 + 6j0 - 300 - j0}{120 + j60 + 300 + j0} = \frac{-180 + j60}{420 + j60} \times \frac{420 - j60}{420 - j60} = 0.4 + j0.2$$

$$|\rho| = \sqrt{0.4^2 + 0.2^2} = 0.446$$

$$VSWR = N = \frac{1 + |\rho|}{1 - |\rho|} = \frac{1 + 0.446}{1 - 0.446} = \frac{1.446}{0.554} = 2.6 \text{ q.e.d}$$

(d)

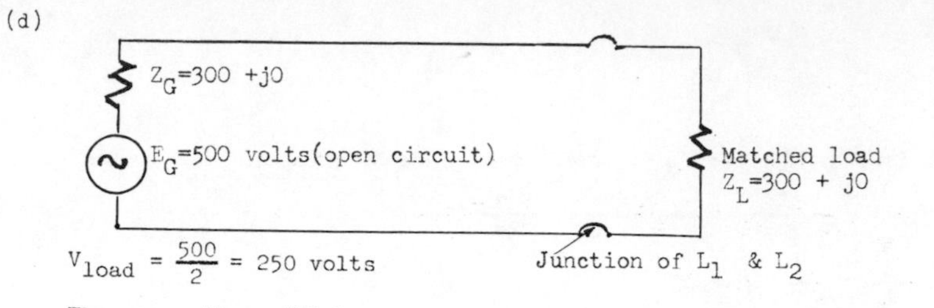

$V_{load} = \frac{500}{2} = 250$ volts

The power then will be supplied to the load under the matched condition:

$P_{load} = \frac{E^2}{R} = \frac{(250)^2}{300} = 208$ watts ANS.

(e) $V_{max} = V_{load}\,(1 + |\rho|) = 250 \times 1.446 = 361$ volts ANS.

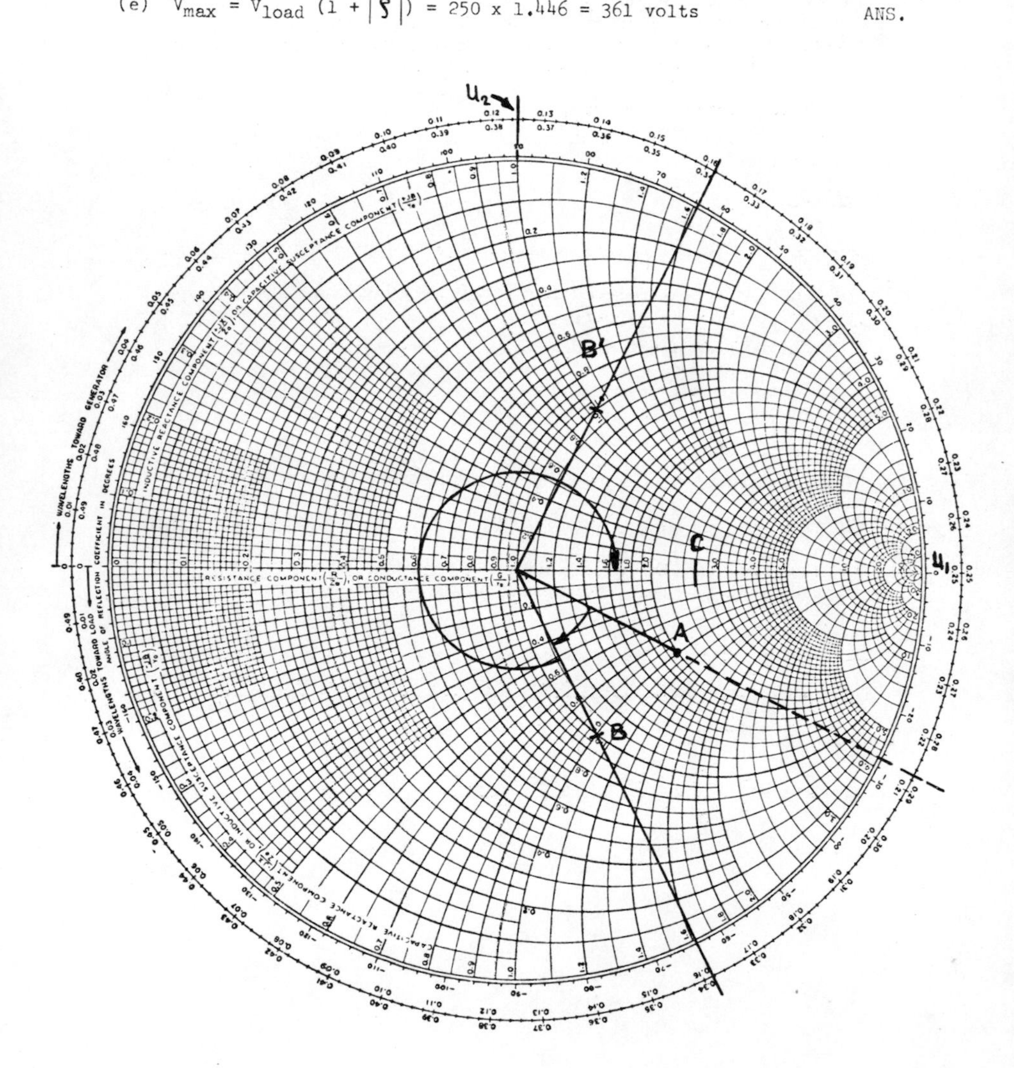

4

MACHINERY

MACHINERY 1.

(1) Wt. 2 A three phase induction motor has the following data on its name plate: 10 hp, 440 volts, 14.1 amps., 60 cps, 1750 rpm. If the motor is operated from a 370-volt, 3-phase, 50-cycle-per-second source, the rotor speed when delivering its name plate rated torque will be approximately:

a. - 1750 rpm g. - 1460 rpm
b. - 1700 rpm h. - 1420 rpm
c. - 1660 rpm i. - 1380 rpm
d. - 1600 rpm j. - 1340 rpm
e. - 1560 rpm k. - 1300 rpm
f. - 1500 rpm l. - None of these

(2) Wt. 2 A 5-hp, 220-volt, d-c shunt motor draws 21.0 amperes at rated load and speed. Manipulating the field control rheostat so as to increase the shunt field circuit resistance by approximately 10% will:

a. - Increase the field circuit current
b. - Decrease the line voltage
c. - Increase the line voltage
d. - Decrease the line current
e. - Decrease the speed
f. - Increase the speed
g. - Cause line current to have lagging power factor
h. - Decrease torque supplied
i. - None of these

(3) <u>Wt. 2</u> An alternator is operating at 1.0 power factor in parallel with a large system. Increasing the field current will:

a. - Decrease the speed
b. - Increase the speed
c. - Increase the line voltage
d. - Decrease the line voltage
e. - Decrease the line current
f. - Increase the line current
g. - Increase the torque supplied
h. - Decrease the torque supplied
i. - None of these

(4) <u>Wt. 2</u> A 5 kva, 2400 - 120/240 volt distribution transformer when given a short-circuit test had 94.2 volts applied with rated current flowing in the short-circuited winding. The per unit impedance of the transformer is approximately:

a. - 0.02
b. - 0.025
c. - 0.03
d. - 0.035
e. - 0.04
f. - 0.045
g. - 0.05
h. - 0.055
i. - 0.06
j. - None of these

(5) <u>Wt. 2</u> In starting a 500 horsepower, 2300 volt, three phase synchronous motor, the field winding is initially short-circuited so as to:

a. - Produce much larger starting torque
b. - Lower induced voltage in field winding
c. - Increase induced voltage in field winding
d. - Provide better flux distribution in the air gap
e. - Lower voltage produced between layers of the field windings
f. - Raise voltage produced between layers of the field windings
g. - Shorten acceleration time
h. - Increase acceleration time
i. - None of these

(1) g. - 1460 rpm approximately.

$$N_{syn} = \frac{f_1 \times 2 \times 60}{p} = \frac{60 \times 2 \times 60}{4} = 1800 \text{ rpm syn. speed}$$

Slip = 1800 - 1750 = 50 rpm

$$s = \frac{1800 - 1750}{1800} = \frac{1}{36}$$

machine has 4 poles

$$N_{syn} \text{ on 50 cps} = \frac{50 \times 2 \times 60}{4} = 1500 \text{ rpm}$$

$$N_{rotor} = 1500 - \frac{1}{36} \times 1500 = 1458.0 \text{ rpm on 50 cps}$$

(2) f. - Increase armature speed.

$$N_{rpm} = \frac{V_T - I_a R_a}{K\phi_f} \quad \text{where } \phi_f \propto I_f$$

If R_f is increased I_f decreases as does ϕ_f.

so N_{rpm} increases.

(3) i. - None of these is exactly correct.

Increasing field current will increase armature current slightly which will increase load on the prime mover. If the prime mover has a governor controlling energy delivered to it, more power will increase line current.

Alternators operating in parallel must have the same frequency and terminal voltage. Increasing the field current will slightly increase the line current which in turn will put a greater load on the prime mover, tending to slow it down. But being locked into the system, the motor-generator cannot slow down, so more power would have to be delivered to the prime mover in order to generate more power. The load delivered by alternators in parallel cannot be changed appreciably by means of the alternator fields. Loads on alternators operating in parallel are changed by shifting the speed-load characteristics of the prime mover. That is, either increasing or decreasing the prime mover power by more or less fuel.

(4) e. - Z per unit = 0.04 approximately.

$I_H = \frac{5000}{2400} = 2.08$ Amperes rated high voltage coil current

$I_L = \frac{5000}{240} = 20.8$ Amperes rated low voltage coil current

$Z_H = \frac{94.2}{2.08} = 45.3$ ohms

$E_{base} = 2400$

$I_{base} = 2.08$ Amperes

$Z_{base} = \frac{2400 \text{ volts}}{2.08 \text{ amperes}} = 1152$ ohms

$Z/\text{unit} = \frac{Z_H}{Z_{base}} = \frac{45.3}{1152} = 0.0392$ ohms

(5) **e. - Lower voltage produced between layers of the field windings. Also, b. would apply.**

The field winding of large synchronous motors has many turns and a large amount of inductance. When voltage is first applied to the stator, high voltages are induced in in the dc rotor field windings if the circuit is open. If the field winding is shorted, short circuit current will flow. The field terminal voltage will be zero and the insulation will not be punctured. If the field is left open, a very high open circuit terminal voltage will develop which may puncture the turn-to-turn insulation, ruining the field winding.

MACHINERY 2.

Subject: Power machinery: Shunt motor

A d-c shunt motor has a name plate rating of 15 hp, 230 volts, 57.1 amp, 1400 rpm. The field circuit has a resistance of 115 ohms and the armature circuit resistance is 0.13 ohm. Neglecting the effect of armature reaction, find:

(1) Wt. 6 The no-load line current

(2) Wt. 4 The no-load speed

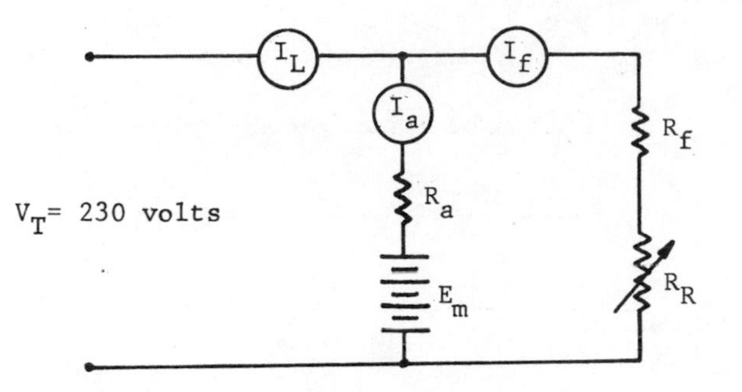

Neglect Armature reaction

$I_a = I_L - I_f = 57.1 - 2 = 55.1$

Input = $V_T I_L$ = 230 x 57.1 =	13,200	watts
$V_T I_f$ = 230 x 2 =	460	"
Armature Input =	12,740	watts
$I_a^2 R_a$ loss = $(55.1)^2(0.13)$ =	396	"
	12,344	watts
Power output = 15 x 746 =	11,200	"
Rotational Losses =	1,144	watts

No Load Losses = $V_T I_a + I_a^2 R_a$

$230\ I_a - I_a^2(0.13) = 1144$

Sign do not appear right.

$0.13\ I_a^2 - 230\ I_a + 1144 = 0$ $\qquad ax^2+bx+c=0$

$$x = \frac{-b \pm \sqrt{b^2-4ac}}{2a}$$

$$I_a = \frac{230 \pm \sqrt{(-230)^2-4(0.13)(1144)}}{2(0.13)}$$

$$= \frac{230 \pm \sqrt{53200 - 595}}{0.26} = \frac{230 \pm \sqrt{52,605}}{0.26} = \frac{230 \pm 229}{0.26}$$

= 3.85 Amps or 1070 Amps

(1) No-load line current: I_a = 3.85 Amps no load

$I_L = I_a + I_f = 3.85 + 2.0 = 5.85$ amps

(2) No-load speed

$$N_{Arm} = \frac{V_T - I_a R_a}{K\phi_f} \quad \text{neglecting armature reaction}$$

$$N_{no\ load} = \frac{230 - 3.85 \times 0.13}{K\phi_f}$$

$$N_{full\ load} = \frac{230 - 55.1 \times 0.13}{K\phi_f} = 1400 \text{ rpm}$$

$$N_{no\ load} = \frac{230 - 0.53}{230 - 7.2} \times 1400 = \underline{1440 \text{ rpm}}$$

MACHINERY 3.

A 50 KVA, 2300/230 volt, 60 cycle transformer is tested in the laboratory so that its characteristics may be determined. The standard test requires an open circuit test and a short circuit test.

Open Circuit Test - Core Loss			Short Circuit Test - Cu Loss		
I	E	W	I	E	W
6.5	230	187	21.7	115	570

REQUIRED:

(a) Calculate the resistance of the windings.

(b) Calculate the copper loss.

(c) Calculate the core loss.

(d) Determine the efficiency of the transformer at full load.

(e) Determine the efficiency of the transformer at half load.

(f) Find the regulation of the transformer for power factor of 1.0.

(g) Find the regulation of the transformer for power factor of 0.8 lag.

(h) Find the regulation of the transformer for power factor of 0.8 lead.

The open circuit test measures core loss with negligible copper loss. The short circuit test measures the copper loss with negligible core loss.

The coils are designed so that

$$I_1^2 R_1 = I_2^2 R_2$$

where I_1 and R_1 are the high voltage coil current and a.c. resistance.

I_2 and R_2 are the low voltage coil current and resistance.

$$\frac{E_1}{E_2} = \frac{N_1}{N_2} = \frac{I_2}{I_1}$$

where $\frac{N_1}{N_2} = a$ (turns ratio)

a) Since

$$W_{Cu\ loss} = I_1^2 R_1 + I_2^2 R_2 = 570 \text{ watts}$$

and

$$I_1^2 R_1 = I_2^2 R_2 \text{ for good transformer design}$$

$$570 = 2 I_1^2 R_1 = 2 I_2^2 R_2$$

$$I_1 = \frac{50,000}{2300} = 21.7 \text{ amps (Rated)}$$

$$I_2 = \frac{50,000}{230} = 217 \text{ amps (Rated)}$$

$$\therefore R_1 = \frac{570}{2(21.7)^2} = 0.605\ \Omega \quad \text{(High Side)}$$

$$R_2 = \frac{570}{2(217)^2} = 0.00605\ \Omega \quad \text{(Low Side)}$$

b) Copper loss $= I_1^2 R_1 + I_2^2 R_2 = 570$ watts

c) Core loss = 187 watts neglecting the no load exciting current copper loss which would amount to $(6.5)^2(0.00605) = 0.255$ watt

d) Full load efficiency

$$= \frac{\text{OUTPUT}}{\text{output} + \text{Cu loss} + \text{CORE loss}}$$

assume Pf = 1

$$\text{Efficiency} = \frac{50{,}000}{50{,}000 + 570 + 187} = \frac{50{,}000}{50{,}757} = 0.985$$

e) Half load efficiency

Assume Pf = 1

$$\text{Efficiency} = \frac{25{,}000}{25{,}000 + \frac{570}{4} + 187} = \frac{25{,}000}{25{,}329.5} = 0.99$$

f) Voltage regulation

$$= \frac{\text{No Load Voltage} - \text{Full Load Voltage}}{\text{Full Load Voltage}}$$

Convert the equivalent transformer circuit, referring it to the low voltage coil. Assume the low voltage coil has constant voltage $E_2 = 230$ volts

$$R_{Equiv._{Low}} = \frac{Power}{I_2^2} = \frac{570}{(217)^2} = 0.0121 \text{ ohm}$$

$R_{Equiv_{Low}}$ $X_{Equiv_{Low}}$ I_2 $E_2 = 230$ volts

$$Z_{Equiv._{High}} = \frac{115}{21.7} = 5.3\ \Omega$$

$$Z_{Equiv._{Low}} = \frac{Z_{Eq.High}}{a^2} = \frac{5.3}{100} = 0.053\ \Omega$$

$$X_{Equiv._{Low}} = \sqrt{Z^2_{Equiv._{Low}} - R^2_{Equiv._{Low}}} = \sqrt{(0.053)^2 - (0.0121)^2}$$

$$= 0.0516 \text{ ohms}$$

Assume rated current $I_2 = 217$ amps

$$\frac{\overline{E_1}}{a} = E_2 + I_2 (R_{equiv._{Low}} + jX_{equiv._{Low}})$$

$$= 230 + 217(0.0121 + j\,0.053)$$

$$= 230 + 2.63 + j\,11.5$$

$$= 232.63 + j\,11.5$$

$$= 232.7 \angle 2.84^\circ$$

Voltage Regulation at $Pf = 1$ $= \frac{232.7 - 230}{230} = \frac{2.70}{230}$

$$= 0.01175 \text{ or } 1.175 \text{ percent}$$

g) Find Voltage regulation for P.f. = 0.8 Lag
Assume rated current $I_2 = 217$ amps

Draw Phasor Diagram

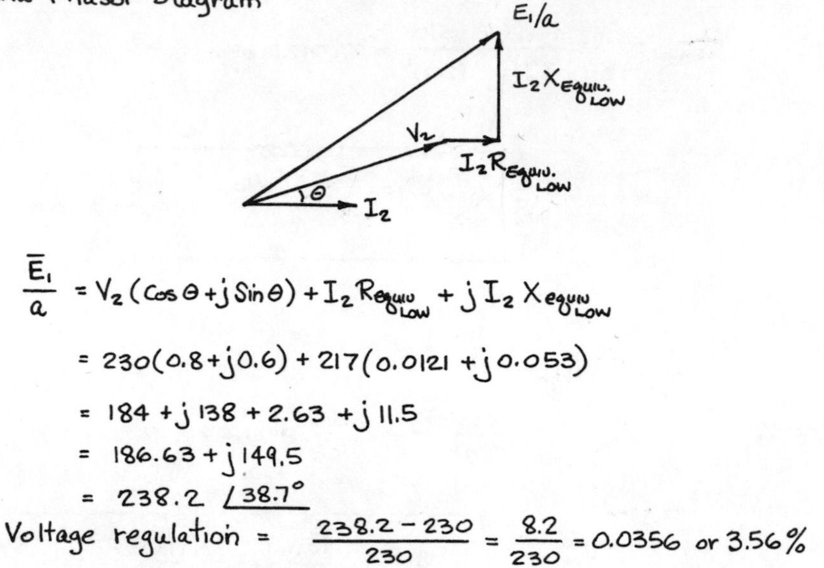

$$\frac{\bar{E}_1}{a} = V_2(\cos\theta + j\sin\theta) + I_2 R_{equiv_{LOW}} + j\, I_2 X_{equiv_{LOW}}$$

$$= 230(0.8 + j0.6) + 217(0.0121 + j0.053)$$

$$= 184 + j\,138 + 2.63 + j\,11.5$$

$$= 186.63 + j\,149.5$$

$$= 238.2 \angle 38.7^\circ$$

$$\text{Voltage regulation} = \frac{238.2 - 230}{230} = \frac{8.2}{230} = 0.0356 \text{ or } 3.56\%$$

h) Voltage Regulation for P.f. = 0.8 Lead

Assume $I_2 = 217$ amperes, rated current

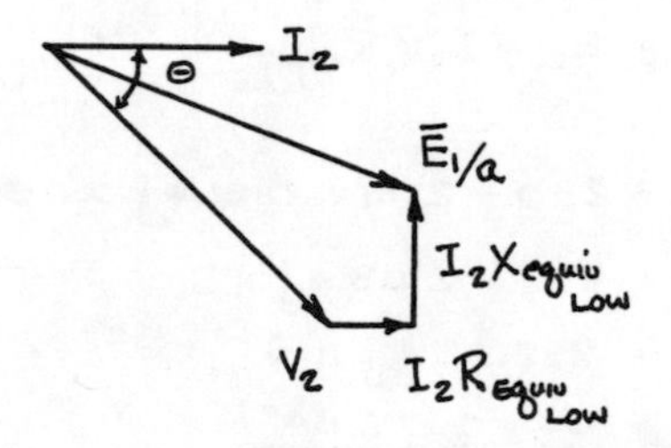

$$\frac{\bar{E}_1}{a} = V_2(\cos\theta - j\sin\theta) + I_2 R_{equiv_{LOW}} + j\, I_2 X_{Equiv_{LOW}}$$

$$= 230(0.8 - j0.6) + 2.63 + j\,11.5$$

$$= 184 - j\,138 + 2.63 + j\,11.5$$

$$= 186.63 - j\,126.5$$

$$= 224 \angle 34.2^\circ$$

$$\text{Voltage regulation} = \frac{224 - 230}{230} = -0.0262 \text{ or } -2.62\%$$

MACHINERY 4.

Subject: Power machinery; synchronous motor torque

A 3-phase, Y-connected, 2300-volt, 50-cps 30-pole, 1000-hp, 1.0-power factor synchronous motor has a synchronous reactance of 3.7 ohms per phase. The motor is supplied from a 3-phase, Y-connected, 1000-kva, 2-pole, 3000-rpm turbine generator whose synchronous reactance is 5.1 ohms per phase.

The field currents of the motor and generator are maintained at those magnitudes which produce rated voltage with 1.0 power factor of the motor at rated motor load. With no other load supplied by the generator the motor mechanical load is gradually increased. Neglecting losses, what is the maximum motor torque in pound-feet that could be produced?

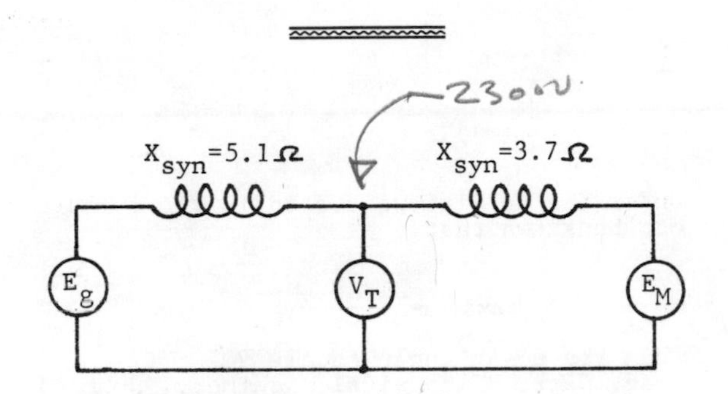

The problem will be solved on a per phase basis. The diagrams are per phase.

$$I_M = \frac{1000 \times 746}{2300\sqrt{3}} = 187 \text{ Amperes Rated Current}$$

$$V_T = \frac{2300}{\sqrt{3}} = 1325 \text{ Volts per phase}$$

$$E_M = V_T - jIX_s = 1325 - j187 \times 3.7 = 1500\underline{/-26.5}$$

$$E_g = V_T + IX_s = 1325 + j187 \times 5.1 = 1630\ \underline{/43.6}$$

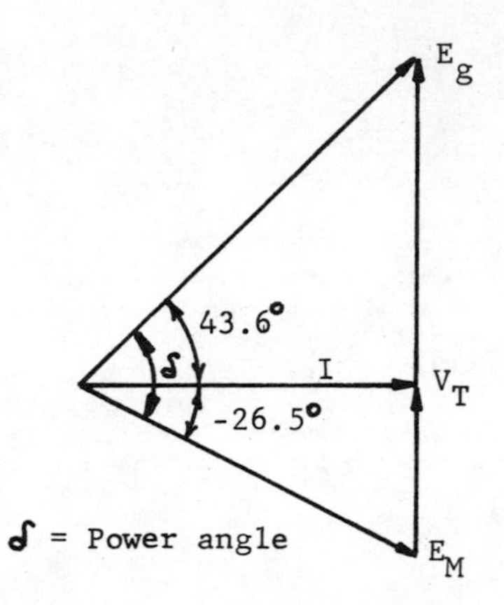

Under the conditions stated in the problem it can be shown that:

$$\text{Power}_{\text{maximum}} = \frac{E_g \cdot E_M}{X_{s_s} + X_{s_m}}$$

when the power angle equals 90°
(See: Carr, "Electrical Machinery", John Wiley and Sons, or Fitzgerald & Kinsley, "Electric Machinery")

$$\text{Power}_{\text{max}} = \frac{1630 \times 1500}{5.1 \times 3.7} = 278{,}000 \text{ Watts/phase}$$

$$3 \text{ Phase Power}_{\text{max}} = 834{,}000 \text{ Watts}$$

$$\text{Torque}_{\text{max}} = \frac{7.04 \times \text{Poles}}{120\ f} \times \text{Watts}$$

$$= \frac{7.04 \times 30}{120 \times 50} \times 834{,}000 = 29{,}300 \text{ Ft-Lbs.}$$

(This neglects losses, armature resistance and saturation.)

MACHINERY 5.

A 100-kva, 11,000/2200-v., 60-cycle, single-phase transformer has an Hysteresis loss of 750 watts, an Eddy-current loss of 225 watts, and a copper loss of 940 watts under the rated conditions of full load. It is desired to export this transformer and to operate it at 45 cps but with the same maximum flux density and the same total loss as at 60 cps.
Calculate the new voltage and kva-rating. Neglect the exciting current.

Solution

Use the expressions for Hysteresis and Eddy-current losses to obtain new values at the new frequency. The new ratings are determined from the values at the new frequency.
The total losses at full load and 60 cps are

$$P_T = P_h + P_e + P_{copper}$$

$$= 750 + 225 + 940 = 1915 \text{ watts}$$

The expressions for core loss are

$$P_h = K_h f B^n_{max}$$

$$P_e = K_e f^2 B^2_{max}$$

Let the primed values indicate values at the new frequency. The modified core losses are

$$P_h' = \frac{45}{60} (750) = 562 \text{ watts}$$

$$P_e' = \left(\frac{45}{60}\right)^2 (225) = 126.5 \text{ watts}$$

Assume the copper loss to be independent of frequency and flux density. Then.

$$P'_{copper} = P_T - P'_h - P'_e$$

$$= 1915 - 562 - 126.5 = 1226.5 \text{ watts}$$

If the voltage is sinusoidal

$$E = 4.44 B_{max} NA \times 10^{-8}$$

then

$$E'_p = \frac{45}{60} (11{,}000) = 8250 \text{v.}$$

The rated primary current at 60 cps is

$$I_p = \frac{kva}{E} = \frac{100{,}000}{11{,}000} = 9.1 \text{ amp}$$

The total equivalent winding resistance referred to the high-voltage is 60 cps is

$$R = \frac{P_{copper}}{I^2_p} = \frac{940}{(9.1)^2} = 11.35 \text{ ohms}$$

The primary current at 45 cps is

$$I' = \sqrt{\frac{P'\ copper}{R}} = \sqrt{\frac{1226.5}{11.35}} = 10.4\ \text{amp}$$

The new kva rating is

$$kva' \ \frac{(8250)(10.4)}{1000} = 86\ \text{kva} \qquad \text{Ans.}$$

The new voltage rating is

8250/1650 Ans.

NOTE: Alternate method to find I'_p.

$$\frac{P_{copper}}{P'_{copper}} = \frac{(9.1)^2 R}{I'_p{}^2 R} = \frac{940}{1,226.5}$$

$$I_p' = \sqrt{\frac{1,226.5}{940}} \times 9.1 = 1.14 \times 9.1 = 10.4\ \text{amp}$$

MACHINERY 6.

Two single phase motors are connected in parallel across a 120 volt, 60 cycle source of supply. Motor "A" is a split-phase induction type and motor "B" is a capacitor type.

Given the following data:

Motor	Horsepower Output	Motor Efficiency	Motor Power Factor
"A"	1/4	0.6	0.7 Lagging
"B"	1/2	0.7	0.95 Leading

REQUIRED:

Wt.		
5	(a)	Find the total power drawn, the combined line current and power factor of the two motors operating in parallel.
5	(b)	Draw a vector diagram showing E_{line}, I_{line}, P_A, P_B, P_{total}. Label all items.

a)

motor	HP_{op}	Eff.	Pf
A	1/4	0.6	0.7 lag
B	1/2	0.7	0.95 lead

Output Motor A = VI Cos Θ × Eff.

In this case

$$HP_{op} = \text{Fraction of load} \times \frac{746 \text{ watts}}{HP} = 0.25\,HP \times \frac{746}{HP} = 186.5 \text{ watts}$$

$$I_{LINE} = \frac{VI \cos\Theta}{V \times Eff. \times Pf} = \frac{0.25 \times 746}{120 \times 0.7 \times 0.6} = 3.7 \text{ amp.}$$

$$\text{Input Motor A} = \frac{\text{Output}}{Eff.} = \frac{186.5}{0.6} = 311 \text{ watts}$$

$$E_{LINE} = 120 \text{ volts}$$

Output Motor B = VI Cos Θ × Eff.

$$= 0.5\,HP \times \frac{746}{HP} = 373 \text{ watts}$$

$$I_{LINE} = \frac{0.5 \times 746}{120 \times 0.95 \times 0.7} = 4.66 \text{ amp.}$$

$$\text{Input Motor B} = \frac{\text{Output}}{Eff.} = \frac{373}{0.7} = 534 \text{ watts}$$

Motor	Pf	VA	Θ	Sin Θ	Input Power
A	0.7 lag	445	45°	0.715	311 watts
B	0.95 lead	562	18.2°	0.3123	534 watts

Total power = Input power A + Input power B

= 311 + 534 = 845 watts

Motor A VARS = 318 lag

Motor B VARS = −176 lead

Total VARS = 142 lag

$$\Theta' = \mathrm{Tan}^{-1}\frac{142}{845} = 9.6^\circ$$

$$\cos\Theta' = \cos(9.6^\circ) = 0.986 \text{ Line Power Factor}$$

b)

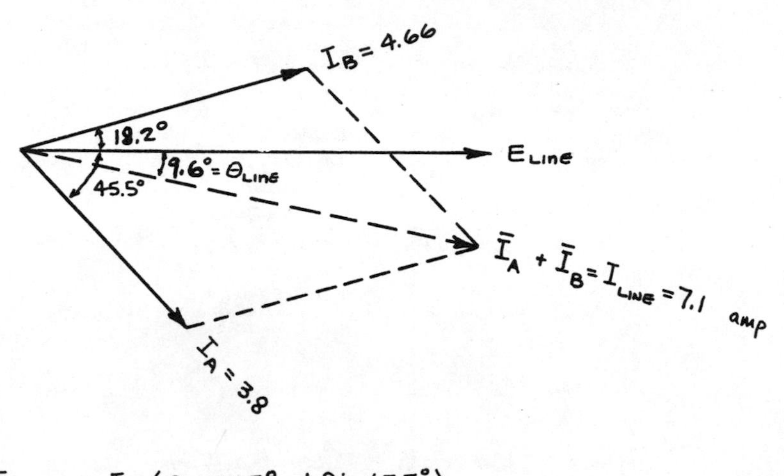

$$\bar{I}_{LINE} = I_A(\cos 45.5^\circ - j\sin 45.5^\circ) + I_B(\cos 18.2^\circ + j\sin 18.2^\circ)$$

$$= 3.7(0.7 - j0.71) + 4.66(0.95 + j0.3123)$$

$$= 2.59 - j2.62 + 4.43 + j1.42 = 7.02 - j1.2$$

$$I_{LINE} = 7.1 \text{ amp.}$$

$P_A = 311$ $P_B = 534$

$$P_{TOTAL} = P_A + P_B = 845 \text{ watts}$$

MACHINERY 7.

In an emergency, a d.c. motor must be used as a generator.

The motor is a cumulative-compound motor; its efficiency is 90%, and its positive and negative terminals are marked.

The cumulative-compound characteristic must be maintained when it is used as a generator, the rotation must be kept in the same direction and the rpm will be the same. The positive terminal must remain the positive terminal in the operation as a generator. The interpoles must aid commutation in both motor and generator mode. The machine is to deliver the same power to the line in the generator-operation as it took from the line in the motor-operation, and the losses in the machine are the same in both cases.

REQUIRED:

(a) Determine if any of the following changes are necessary, and do enough calculations to verify your answer:

1. Should the shunt-winding connections be changed?
2. Should the compound-winding connections be changed?
3. Should the interpole-winding connections be changed?
4. Should the field resistance be changed?

(b) Has the load on the mechanical clutch of the machine increased or decreased when the operation changed from motor to generator if it was running at full capacity as a motor?

(c) How much does the electrical power transferred at the line connection increase or decrease if the same mechanical torque is maintained on the shaft?

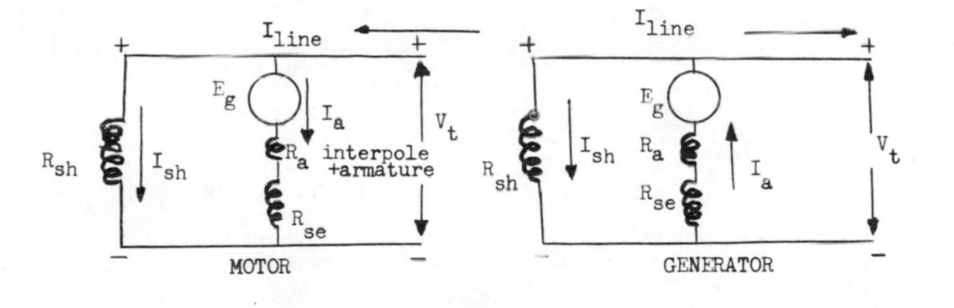

(a) 1. No, the shunt winding should not be changed. The problem clearly states that direction of rotation and polarity (positive terminal) stay the same. Therefore regardless of whether we have a motor or generator operation, the direction of the current stays unchanged. The only draw-back is the subtractive MMF of shunt and series field winding, giving a differential compounding. This matter is brought up in the next point.

2. Yes, the compound (series) winding connections should be changed, since the armature current during generator operations is reversed and now opposes (differentially compounded) the shunt field. To still aid (cumulatively compounded) the shunt field, the compound (series) winding should be reversed.
3. No, the interpole winding connections should not be changed. Interpoles or commutating poles are narrow laminated auxiliary poles placed midway between the main poles and the plane of commutation. These interpoles are in series with the armature and are wound to oppose and nullify the armature reaction in the commutating plane. This prevents sparking that might cause flashover and also reduces iron losses in the armature teeth.
   Changing from motor to generator action, the polarity of the commutating pole automatically changes, with the change of the armature reaction MMF. Therefore, commutation in interpole machines is not affected by a change from motor to generator operation or a change in the direction of rotation.
4. Yes, the field (shunt) resistance should be changed. Since the generator voltage has to be larger than the terminal voltage due to the ohmic voltage drop in the armature winding, the flux has to be increased ($E_G = K .\emptyset.$ rpm). Increased $\emptyset$ means increased field current. Therefore, the shunt field resistance should be decreased.

(b) The load on the mechanical clutch of the machine has increased.

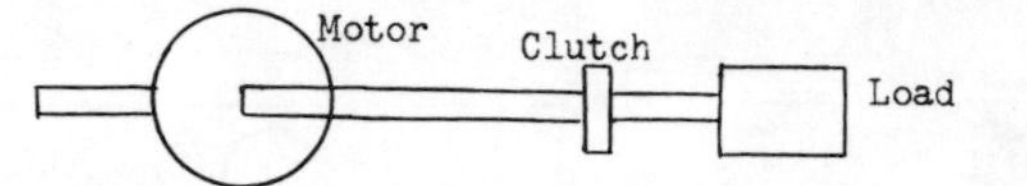

Rated $Power_{input}$ = Losses + Load on clutch (for motor)

Rated $Power_{output}$ + Losses = Load on clutch (for generator)

Since losses remain the same in either mode of operation, the load on the clutch increases when machine is operated as a generator.

(c) Using the line connections as reference point, due to internal losses, the power delivered to the motor shaft is 90% of the reference or input power.
Assuming that the same mechanical torque is maintained, above shaft power is now the input (prime mover) of the generator operation. Therefore, using the same 90% efficiency of the machine, the generator output referred to the original electrical power transferred at the line connections has decreased to: 0.90 x 0.90 = 0.81 or 81%

Reference: Electrical Circuits and Machines, Chapter XIII, by Robertson and Black.

MACHINERY 8.

Two Y connected, 50° rise induction motors are fed by a 4160 V, line-to-line, 3-phase 60 Hz motor-control center 20 feet away. Motor #1 drives a 600 HP compressor. The efficiency of the motor is 90%, and its power factor is 0.5. Instruments of motor #2 indicate 1730 kw, 277 amps.

REQUIRED:

(a) Show the phasor-diagram of the loads, kw and kva.
(b) Determine the capacity in microfarads of a wye-connected capacitor bank that is required to correct the power factor of the total load to 0.966.
(c) If a synchronous motor is installed in place of motor #2 and used instead of the capacitor bank to achieve the same over-all power factor (0.966), what must its power factor be?
(d) Determine the feeder size (copper) and the rating of the control center circuit breaker, or fuse, that must be used for each of the following conditions, and indicate what section of the Electrical Safety Orders is applicable: (1) The 2 induction motors. (2) The 2 motors with the capacitor bank connected. (3) 1 induction motor and 1 synchronous motor.

(a) Motor #1 = $\frac{600}{0.90} \times 0.746 = 497 \text{ KW} \cong 500 \text{ KW}$

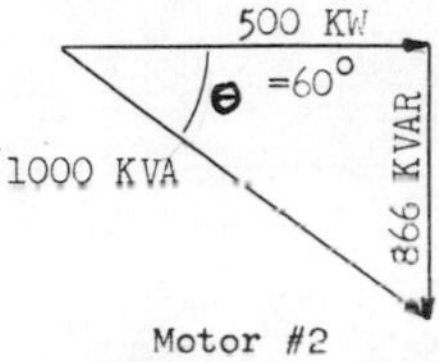

500 KW; θ =60°; 1000 KVA; 866 KVAR

pf = 0.5, thus θ = 60°

$\text{KVA} = \frac{500 \text{ KW}}{\cos 60°} = \frac{500}{0.5} = 1{,}000$

$\text{KVAR} = 1{,}000 \times \sin 60° = 1{,}000 \times 0.866 = 866$

Motor #2

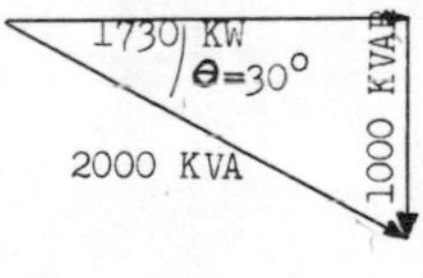

1730 KW; θ=30°; 2000 KVA; 1000 KVAR

$\text{KVA} = \sqrt{3} \times 4{,}160 \times 277 = 1994 \cong 2{,}000$

$\text{pf} = \cos \theta = \frac{1{,}730}{2{,}000} = 0.866; \ \theta = 30°$

$\text{KVAR} = 2{,}000 \times \sin 30° = 2{,}000 \times 0.5 = 1{,}000$

(b) Total load of motors #1 and #2

$\text{KW} = 500 + 1{,}730 = 2{,}230$
$\text{KVAR} = 866 + 1{,}000 = 1{,}866$
$\text{KVA} = \sqrt{2{,}230^2 + 1{,}866^2} = \sqrt{8.45 \times 10^6} = 2{,}907$

Actual combined pf $= \cos \theta = \frac{2{,}230}{2{,}907} = 0.767$ lag; $\theta = 40°$.

Desired combined pf = 0.966 lag; $\theta = 15°$

$KVA\ new = \frac{2,230}{0.966} = 2,308$

BC = Required leading KVAR = $1,866 - 2,230 \tan 15°$
$= 1,866 - 2,230 \times 0.268 = 1,866 - 598 = 1,268$

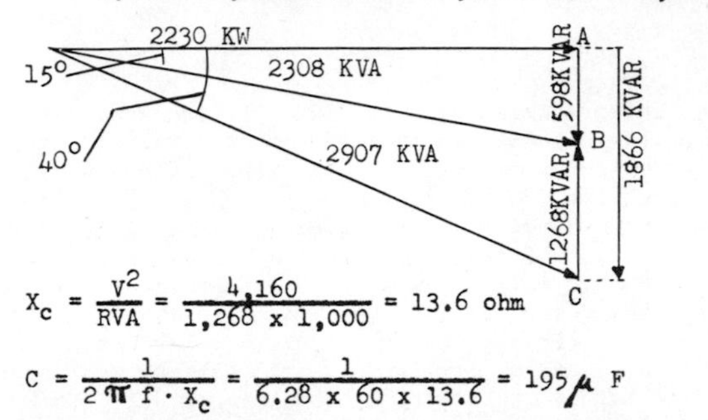

$X_c = \frac{V^2}{RVA} = \frac{4,160}{1,268 \times 1,000} = 13.6$ ohm

$C = \frac{1}{2\pi f \cdot X_c} = \frac{1}{6.28 \times 60 \times 13.6} = 195 \mu F$ ANS.

(c) Assuming that synchronous motor has same efficiency as motor #2 which it replaces,

KVAR syn. motor = KVAR motor #1 - KVAR desired = AC - AB
= 866 - 598 = 268

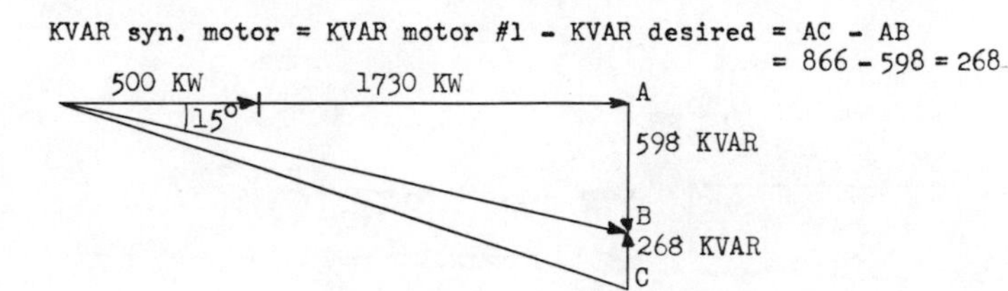

The synchronous motor alone:

1750 KVA
268KVAR
α
1730 KW

$\tan \alpha = \frac{268}{1,730} = 0.155; \ \alpha = 9°$

$KVA = \frac{1,730}{\cos} = \frac{1,730}{0.988} = 1,750$

$pf = \cos 9° = 0.988$ ANS.

(d)
(1) Two induction motors
A. For feeder size, from ESO 2395 (a), p. 261

$1.25 \times I_{motor\ \#2} + I_{motor\ \#1} = 1.25 \frac{2,000,000}{\sqrt{3} \times 4,160} + \frac{1,000,000}{\sqrt{3} \times 4,160}$

$= 1.25 \times 279 + 139 = 348 + 139 = 487$ amps

From ESO Article 38 Table 1A p. 358.22 for non-continuous loads (i.e., for other than air conditioning motors) we select a 750 MCM feeder. ANS.

B. For overcurrent protection of motors from ESO 2396 (a), p. 262 and Article 38, Table 10, p. 358.22, column 8 (we assume squirrel cage motor without code letters, since none are given, and more than 30 amperes):

$I_{motor\ \#2}$ = 278 (the largest current) has breaker or fuse rating of 600 amps.

Total rating = 600 + $I_{motor\ \#1}$ = 600 + 139 = 739 amps

Since this is not a standard size, we use the next higher value, or 800 amps. ANS.

(2) The 2 motors with the capacitor bank.

A. $I_{total} = \frac{2,308,000}{\sqrt{3} \times 4,160} = 321$ amps

We assume that the capacitor is located at the motor.

$I_{motor\ \#1} = 321 \times \frac{500}{2,230} = 72$

$I_{motor\ \#2} = 321 - 72 = 249$

To select feeder:
$1.25 \times 249 + 72 = 311 + 72 = 383$ amps

From Table 1A we select size 500 MCM or 2 sets of 250 MCM feeders. ANS.

B. $I_{motor\ \#2}$ = 249 ......500 amp breaker

$I_{motor\ \#1}$ = 72 amps

Total rating = 500 + 72 = 572 amps. Select nearest size of 600 amps. ANS.

(3)

A. $I_{syn\ motor} = \frac{1,750,000}{\sqrt{3} \times 4,160} = 243$ amps

$I_{motor\ \#1}$ = 72 amps

These values are very close to above case (2)A. We select same size 500 MCM or 2 sets of 250 MCM feeders. ANS.

B. Breaker rating is 600 amps; see above case (2)B. ANS.

5

POWER DISTRIBUTION

PROBLEM 1.

A 345 KV power transmission line has two bundled conductors per phase, spaced 18 inches horizontally. The conductor used in the bundle has a self GMD of 0.0403 feet and the phases are spaced horizontally 15½ feet apart.

REQUIRED:

Determine the following:
(a) The self GMD of the bundled conductors.
(b) The mutual GMD of the line.
(c) The inductive reactance per phase per mile.

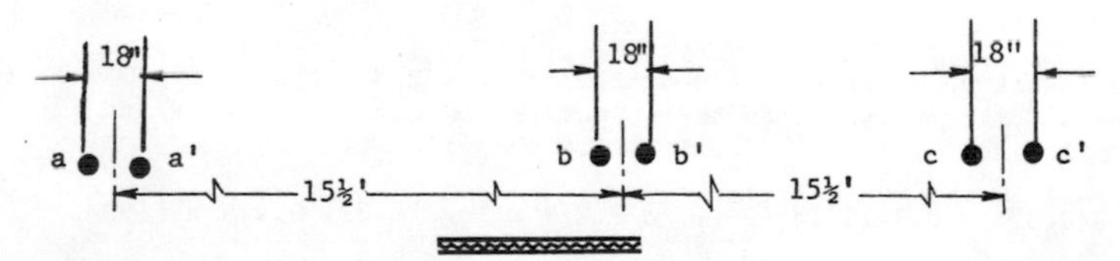

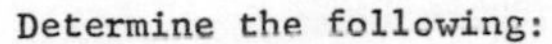

(a)
Often self GMD of a bundled or composite conductor is called "geometric mean radius", or GMR. Self GMD may be denoted as D_s. This term includes the distances of a strand or conductor from all other strands within the same bundle plus the "distance of the strand from himself" or the self GMR of the strand.

In above line configuration we have two strands per bundle, thus we have four distances: $D_{aa'}$, $D_{a'a}$, D_{aa}, $D_{a'a'}$. The self GMR of a single strand is less than the actual physical radius (R x 0.7788). This reduced radius is the above given self GMD of 0.0403 ft. and is available from tables. Converting all distances into feet, we obtain for one bundle:

$$D_s = D_{sa} = \sqrt[4]{D_{aa'} \times D_{a'a} \times D_{aa} \times D_{a'a'}} = D_{sb} = D_{sc}$$

Note: We extract the fourth root as we have four distances under the radical

$$D_s = \sqrt[4]{(0.0403)^2 \times \frac{18}{12}^2} = \sqrt[4]{0.00366} = \frac{\log 0.00366}{4} =$$

$$= \frac{\log 3.66 - 4}{4} = \frac{0.564 - 3}{4} = \frac{1.564 - 4}{4} = 0.391-1$$

$$= 0.246 \text{ ft.} \quad \text{ANS.}$$

(b)
The mutual GMD of the line or D_{eq} is the geometric mean of all mutual GMD values outside the bundles, i.e. between the three phases.

$$D_{ab} = D_{bc} = \sqrt[4]{(15.5)^2 \times 17.0 \times 14.0} \text{ where } ab = 15.5'; \; a'b' = 15.5'$$

$$ab' = 15.5 + 2 \times \frac{18}{12} \times \frac{1}{2} = 17.0'$$
$$a'b = 15.5 - 2 \times \frac{18}{12} \times \frac{1}{2} = 14.0'$$

$$= \sqrt[4]{57{,}180} = \frac{\log 57{,}180}{4} = \frac{4.758}{4} = 1.19 = 15.5 \text{ ft.}$$

$$D_{ac} = \sqrt[4]{(15.5 \times 2)^2 \times 32.5 \times 29.5} = \sqrt[4]{921{,}400} = \frac{\log 921{,}400}{4}$$

$$= \frac{5.966}{4} = 1.492 = 31 \text{ ft.}$$

$$D_{eq} = \sqrt[3]{D_{ab} \times D_{bc} \times D_{ac}} = \sqrt[3]{15.5 \times 15.5 \times 31.0} = 19.5 \text{ ft.} \quad \text{ANS.}$$

(c)
The inductive reactance in ohm/mile or $X_L = 2\pi f \times 10^{-3} \times 0.7411 \log \frac{D_{eq}}{D_s}$.
The 10^{-3} factor is needed to convert the inductance L, obtained in mh/mile into h/mile to finally yield ohm/mile.

$$X_L = 0.377 \times 0.7411 \log \frac{19.5}{0.246} = 0.279 \times \log 79.3 = 0.279 \times 1.90 =$$

$$= 0.530 \text{ ohm/mile} \quad \text{ANS.}$$

Reference: William D. Stevenson, Jr. "Elements of Power System Analysis", McGraw Hill. p. 30-38.

PROBLEM 2.

Power for a remote building on an industrial site is supplied through an existing buried cable from a fixed voltage 60-cycle supply.

The load in the remote building consists of lighting and induction motors. During periods of peak demand, when the cable is carrying approximately its rated current, the resulting steady-state load voltage is well below the desired value because of the characteristics of the load. A small amount of additional constant-speed motor load is anticipated in the near future.

REQUIRED:

What equipment can be installed at the building to improve the present situation and to permit the additional load? Explain how the equipment you recommend will improve the situation; the use of phasor diagrams is suggested.

The crux of the whole problem is the large portion of induction motors. The power factor of induction motors at rated load are typically from 0.70 to 0.90 with some groupings of motors resulting in even lower power factors.

For a power factor of 0.8, a motor drawing 225 KVA of power will utilize only 180 KW

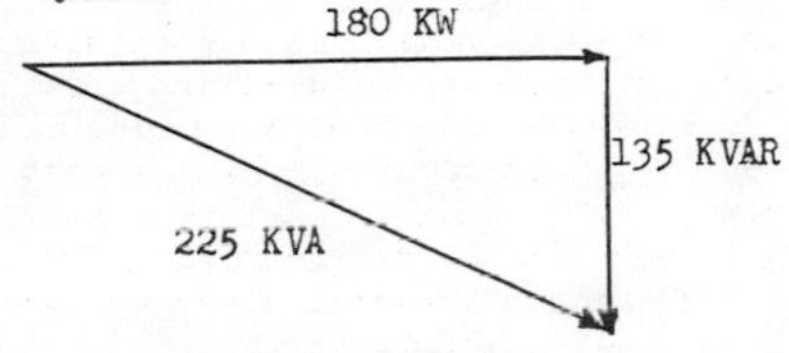

KW = Real power

KVAR = Reactive power

KVA = Apparent power

If the KVA drawn in this case were equal to the real power required (KW) a 20% reduction in current would result. The reduction in current with present load would reduce the voltage drop, thus improve the voltage at the load.

Fluorescent lighting with capacitors usually has power factors from 0.95 to 0.97, therefore are not practical to try to improve the power factor any higher.

To improve the power factor with the existing loads, capacitors should be applied. They have the characteristic of a leading power factor whereas induction motors have a lagging power factor. By adding capacitors their leading KVAR cancels out the equivalent amount of lagging KVAR, i.e.:

180 KW

135 KVAR resulting from a group of induction motors

135 KVAR of capacitors

Above power factor correction results in 180 KVA = 180 KW or power factor = 1.0.

Another method of improving power factor is to add synchronous motors for the additional motor requirements. The synchronous motor acts like a capacitor producing a leading power factor and leading KVAR.

Normally capacitors are the most effective in reducing system costs when located near the devices with low power factor. Here we are primarily concerned with the feeder to the building, but if there are any large induction motors or a grouping of motors it would minimize local branch circuit voltage drop as well as the feeder, if the capacitors were located near the source of the low power factor (pf).

An economic study of the situation should be made. Data from recording pf meters and KW meters should be gathered from as many places as feasible on feeder and branch circuits. Then a comparison of how much and where the capacitors should be installed. The installation of the synchronous motors vs. induction motors with capacitors should be evaluated.

With the additional load the voltage drop in the existing feeder may be too much even with unity pf. Then consideration should be given to using boost transformer. It may be well that a combination of boosting and capacitor and synchronous motor will be the most economical solution. Boost transformers are much less in cost than regular transformers as they are just an auto-transformer. Improving the power factor beyond a certain point increases the cost disproportionately to the gain obtained, thus all alternates should be weighed in making the ultimate decision. Using a boost transformer alone may mean that at light load an overvoltage may result which could be undesirable.

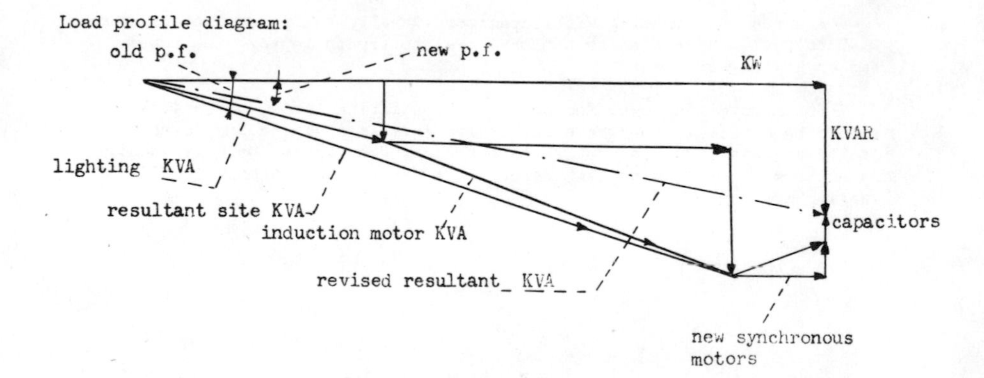

PROBLEM 3. ?

In the diagram below, determine the fault currents at point "F" for the following conditions:

Wt.
4 (a) A 3-phase fault.

6 (b) A single line to ground fault.

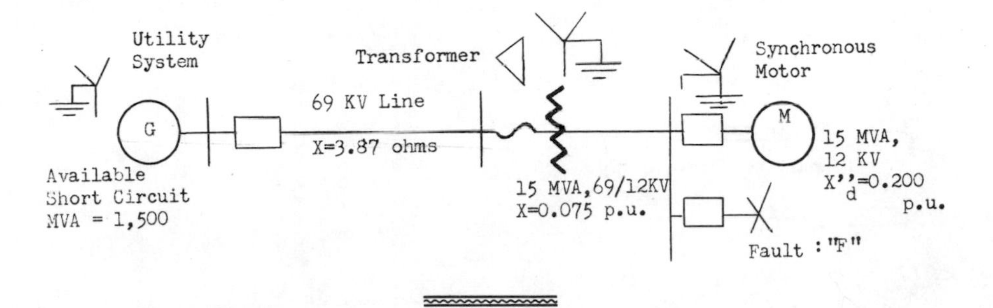

KVA_{base} = 150 MVA (This value was selected to be a practical base between the two given MVA values)

KV_{base} = 69 and 12 respectively

$$Z_{base\ 69} = \frac{(KV)^2 \times 1000}{KVA_{base}} = \frac{(69)^2 \times 1000}{150{,}000} = 31.8 \text{ ohms}$$

$$Z_{base\ 12} = \frac{(12)^2 \times 1000}{150{,}000} = 0.96 \text{ ohms}$$

$$I_{base\ 69} = \frac{KVA_{base}}{\sqrt{3}\, KV_{base}} = \frac{150{,}000}{\sqrt{3} \times 69} = 1{,}250 \text{ amps}$$

$$I_{base\ 12} = \frac{150{,}000}{\sqrt{3} \times 12} = 7{,}230 \text{ amps}$$

$$Z_{line\ p.u.} = \frac{Z_{rated\ ohms}}{Z_{base}} = j\frac{3.87}{31.8} = j0.121 \text{ p.u.}$$

$$Z_{trans\ p.u.} = Z_{rated\ p.u.} \frac{KVA_{base}}{KVA_{rated}} = j0.075 \frac{150{,}000}{15{,}000} = j0.750 \text{ p.u.}$$

$$Z_{utility} = 1.0 \frac{KVA_{base}}{KVA_{sh.ckt.}} = j \times 1.0 \frac{150{,}000}{1{,}500{,}000} = j0.100 \text{ p.u.}$$

$$Z_{motor\ dl''} = j0.200 \frac{150{,}000}{15{,}000} = j2.00 \text{ p.u.}$$

(a) Three phase fault

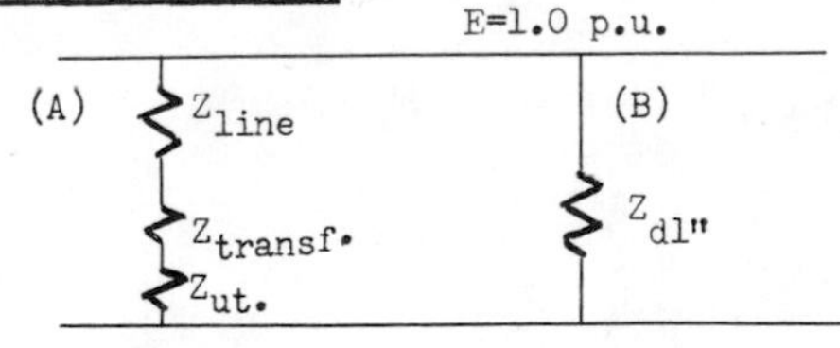

$Z_A = Z_{line} + Z_{transf} + Z_{ut} = j0.121 + j0.750 + j0.100 = j0.971$

$Z_B = Z_{d1''} = j0.200$

$$Z_{eq} = \frac{1}{j0.971} + \frac{1}{j0.200} = \frac{j0.971 \times j0.200}{j0.971 + j0.200} = j0.655$$

$$I_{fault} = \frac{E}{Z_{eq}} = \frac{1.0}{j0.655} = -j1.525 \text{ amps}$$

I_{fault} 3 ph at 12 KV $= I_{fault\ p.u.} \times I_{base}$ 12 KV

$= -\ j1.525 \times 7{,}230 = -\ j11{,}000$ amps ANS.

(b) Single phase fault

Positive sequence impedance diagram:

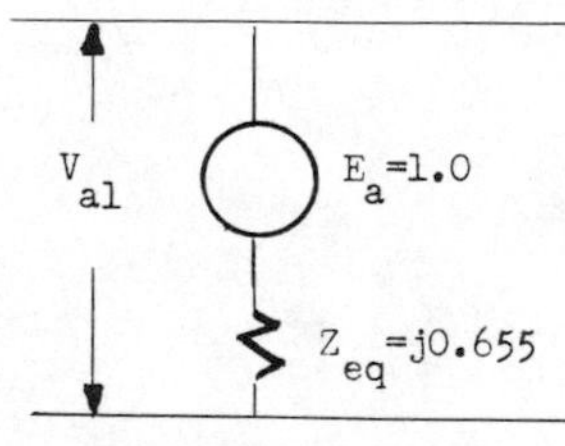

Negative sequence impedance diagram:

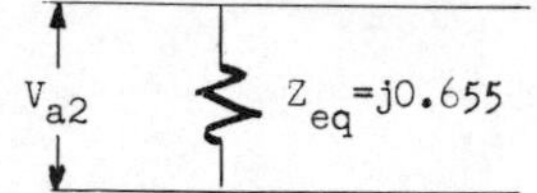

Zero sequence impedance diagram:

69 KV system zero sequence fault currents are isolated from fault "F" by ΔY transformer

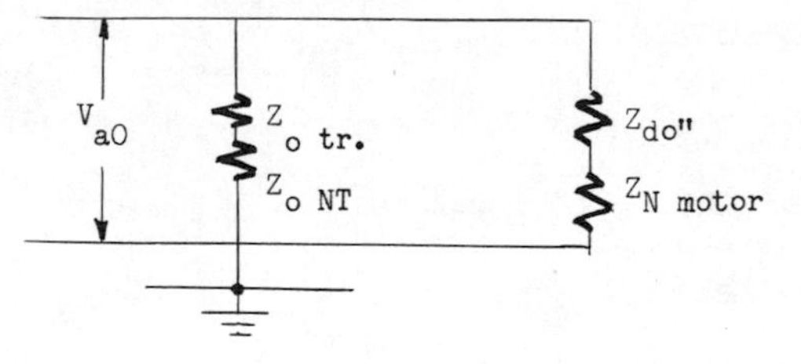

$Z_{o\ tr} = j0.750$

$Z_{motor\ do''} = \frac{1}{2} Z_{motor\ d1''}$ (assumed) $= j1.000$

$Z_{o\ NT} = 3Z_{N\ motor} = 0$ .... directly connected neutrals

$$Z_{o\ eq} = \frac{j0.750 \times j1.000}{j0.750 + j1.000} = j0.428$$

Sequence Network

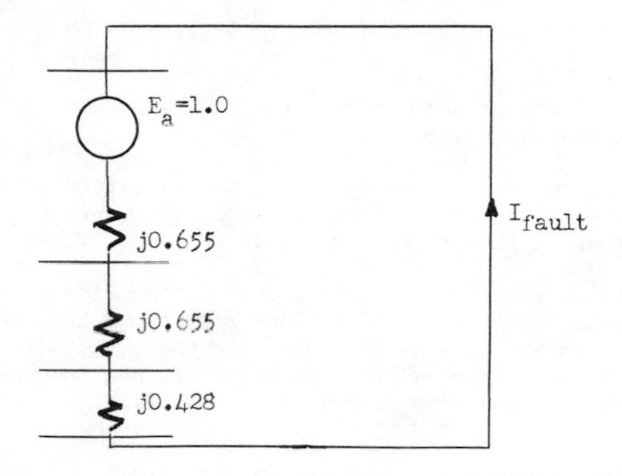

$$I_{a1} = I_{a2} = I_{ao} = \frac{E_a}{Z_1 + Z_2 + Z_o + 3Z_N + 3Z_{fault}}$$

Where $Z_N = 0$ and $Z_{fault} = 0$

thus:

$$I_{a1} = I_{a2} = I_{ao} = \frac{1.0}{j(0.655 + 0.655 + 0.428)} = -j0.580$$

$I_{fault\ p.u.} = (I_{a1} + I_{a2} + I_{ao}) = 3\ (-j0.580) = -j1.740$

$I_{fault} = I_{fault\ p.u.} \times I_{base\ at\ 12\ KV}$

$= -j1.740 \times 7{,}230 = -j12{,}500$ amps ANS.

PROBLEM 4.

An emergency 120/208-volt, 3-phase, 4-wire, 60-cps generator supplies an external circuit.

The load on the external circuit consists of 9000 watts of incandescent lights connected between line and neutral and evenly distributed among the 3 phases, and a 10-HP, 3-phase air conditioner motor of 83% efficiency and 0.707 P.F.

REQUIRED:

Wt.		
5	(a)	Show the phasor-diagram of the currents and voltages on the load-side of the generator.
3	(b)	Determine the microfarads of the capacitor to be connected to the generator in order to reduce the generator load-current to 105% of that which would flow if the P.F. = 1.
2	(c)	Determine the size of the conduit and wire to be used as a feeder and the required fuse size to protect the feeder when it is run between the generator and its distribution board in the next room, if the total load is continuous, and the capacitor calculated in (b) is connected.

(a) Designating: I_m = Motor current; I_{mR} = Real component of I_m; I_{mQ} = Quadrature component of I_m; I_i = Incandescent lights current; I_1 = Total load current; I_1' = Corrected total load current; I_c = Capacitor current,

We obtain as follows:

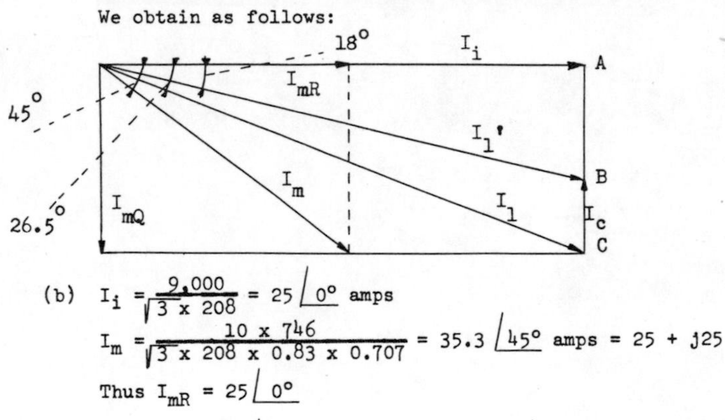

(b) $I_i = \dfrac{9,000}{\sqrt{3} \times 208} = 25\angle 0°$ amps

$I_m = \dfrac{10 \times 746}{\sqrt{3} \times 208 \times 0.83 \times 0.707} = 35.3\angle 45°$ amps $= 25 + j25$

Thus $I_{mR} = 25\angle 0°$

$I_{mQ} = 25\angle -90°$

$I_1 = 25 + 25 + j25 = 50 + j25 = 56\angle 26.5°$ amps

If p.f. = 1, then I_1 reduces to the real component of above current i.e. $50\angle 0°$. Therefore, we have to reduce $I_1 = 56\angle 26.5°$ to 105% of 50 amps = $I_1' = 52.50$ amps at an angle to be determined.

From above vector diagram I_1' (52.50) ends in point B

$AC = 25$

$AB = \sqrt{52.5^2 - 50.0^2} = 16.5$

$BC = 25 - 16.5 = 8.5$ amps $= I_c$

Then: $X_c/\text{phase} = \frac{E}{\sqrt{3} \times I_c} = \frac{208}{\sqrt{3} \times 8.5} = 14.1$ ohms

If $X_c = \frac{1}{2\pi fC}$

Then $C = \frac{1}{377 \times 14.1} = 18.7\ \mu F/ph.$

$C_{total} = 3 \times 18.7 = 56.1\ \mu F$ ANS.

(c) $I_1' = (25 + 25) + j16.5 = 52.50\angle 18°$ amps

$I_{m\ new} = 25 + j16.5 = 30\angle 34°$ amps

Rating of wire size:

$125\%\ I_{m\ new} + I_i = 1.25 \times 30 + 25 = 62.5$ amps

From copper wire tables (ESO, Section 358.22 Table 1A) for 70 amps continuous current we obtain Size No. 4 ANS.

From over current protection table (ESO, Section 358.30 **Article** 38) at 30 amps (no 125% factor is needed) + 25 = 55 amps we obtain a fuse size of 120 amps ANS.

From conduit table (ESO, Section 358.26 Article 38, Table 4) for 4 wires and 55 amps we obtain a conduit size of 1 1/4" ANS.

PROBLEM 5.

There are presently available several different types of systems for use as 69 KV underground transmission lines. These systems are:

LPOF Single conductor cable

LPOF Three conductor cable

Solid type single conductor cable

MPGF

HPOF

HPGF

REQUIRED:

(a) Briefly describe each system, and list conditions that would favor the use of one system over another. Justify your answers.

(b) Which system is the most reliable?

(c) Which system is the least reliable?

(d) Which system costs the least (to install)?

(e) Which system probably costs the most for short runs?

(a) 1. Low Pressure Oil-Filled Single Conductor Cable (LPOF).

Metallic oil channels are provided in place of filler material between the conductor and the load sheath. The oil-filled cable was developed to overcome the migration of compound and resultant voids inherent in solid cables by maintaining a static head of oil on the cable at all times. This is done by means of reservoirs. The mechanical strength of the lead sheath limits the static head permissible from 30 to 40 feet. Also, because of resistance to flow of oil, there are certain limitations on the length of a section of cable. Special stop joints are required between sections supplied by different reservoirs, and special terminating equipment must be used. When filling the oil system, special care must be exercised to thoroughly dry and de-gasify the oil.

2. Low Pressure Oil-Filled Three Conductor Cable (LPOF).

This system is similar to that of (a)1. above, with the exception that the cable contains 3 conductors instead of 1, and is less expensive. Three conductor oil-filled cable may be round or sector type and may be used on voltages from 20-69 KV between phases.

3. Solid Type Single Conductor Cable.

Impregnated paper insulation is wrapped around a single solid conductor cable. It cannot dissipate heat as well as oil- or gas-filled cables and therefore has a much shorter life. However, the cable is relatively inexpensive. Single conductor, solid type of paper - insulated cable may be used up to 70 KV, whereas the 3-conductor cables are rarely used above 35 KV because of excessive size and weight in the higher voltages.

4. Medium Pressure Gas-Filled Cables (MPGF).

   This cable is filled with inert gas (hydrogen) under a pressure of about 200 pounds per square inch. The inert gas fills the space between the paper insulted cable and the lead sheath. This cable has the advantage of being cooled without maintaining a reservoir. Further, the inert gas encounters very little resistance to flow.

5. High Pressure Oil-Filled Cables (HPOF).

   This system is similar to that of (a)2. above, with the advantage that oil under greater pressure has higher dielectric strength. However, the mechanical problems are increased. Special stop joints and terminating equipment are required to withstand the high pressures.

6. High Pressure Gas-Filled Cables (HPGF).

   This system is similar to that of (a)4. above, with the advantage that the inert gas under greater pressure has higher dielectric strength. However, the mechanical problems are increased.

(b) In general, the high pressure gas-filled cable system is the most reliable. It has comparatively few sources of mechanical problems.

(c) In general, the solid type single conductor system is the least reliable. Because of its poorer ability to dissipate heat, it is subject to higher voltage stresses, resulting in a shorter life.

(d) The solid type single conductor system would cost the least to install.

(e) In general, the high pressure gas-filled system would be the most expensive for short runs.

Reference: Standard Handbook for Electrical Engineers by A. E. Knowlton, Editor in Chief, Section 13, paras 210 to 220.

6

ELECTRONICS & COMMUNICATIONS

PROBLEM 1.

Calculate the power gain of the two-stage amplifier shown below. Show the gain of the two individual stages, the interstage losses and the pre-first stage losses. The ground-based "h" parameters for the transistors used in both stages are:

h_{ib} = 50 ohms; h_{rb} = 5×10^{-4}; h_{fb} = -0.97

and h_{ob} = 10^{-6} mho

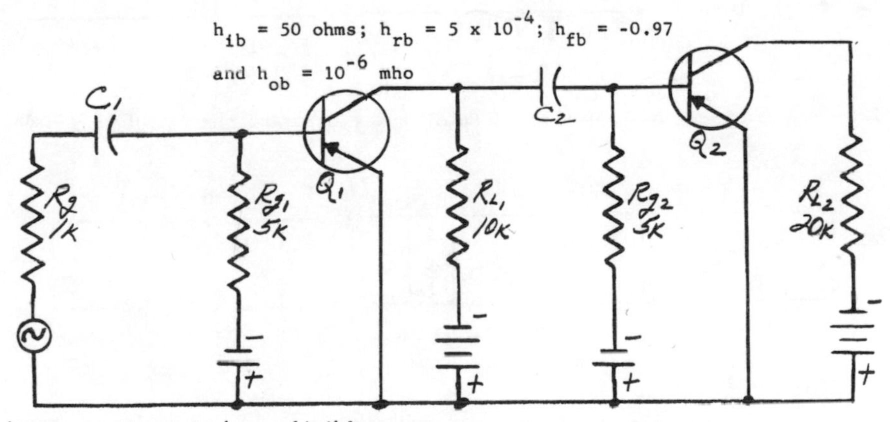

Assume reactance to be negligible.

Solution based upon the following assumptions:

1. Mid-band frequency -- reactance of capacitors is negligible.
2. Power gain defined as power delivered to load divided by power delivered from source (this is power delivered to R_{L2} divided by power delivered the voltage node at the junction of C_1 and R_{g1}).
3. Power losses are defined as:
   a) Pre-first stage -- power lost in bias resistor R_{g1}.
   b) Inter-stage -- power lost in R_{L1} and R_{g2}.
   c) Power losses -- at signal frequency only, d.c. bias losses <u>not</u> considered.

Parameters are given in common base configuration; since transistors are operated in common emitter orientation, the parameters must be converted to common emitter form. Also, gain and impedance equations must either be derived or found in a transistor handbook.

Handbook conversion tables (from Fig. 4.11, G.E. Transistor Manual, 6th ed.) give the following relationships:

$$h_{ie} = \frac{h_{ib}}{1+h_{fb}} = 1670\ \Omega \qquad h_{re} = \frac{h_{ib}h_{ob}}{1+h_{fd}} = 11.7\times10^{-4}$$

$$h_{fe} = \frac{-h_{fd}}{1+h_{fb}} = 32 \qquad h_{oe} = \frac{h_{ob}}{1+h_{fd}} = 3.3\times10^{-5}\ \mho$$

And (from Fig. 4.14):

$i_1 \rightarrow$ | v_1 | h Parameter Network | v_2 | $\leftarrow i_2$ | $R_L = \frac{1}{G_L}$

$$\text{I.}\quad Z_{in} \triangleq \frac{v_1}{i_1} = \frac{h_{ie}h_{oe} - h_{fe}h_{re} + h_{ie}G_L}{h_{oe}+G_L} = h_{ie} - \frac{h_{re}h_{fe}}{h_{oe}+G_L}$$

$$\text{II.}\quad A_v \triangleq \frac{v_2}{v_1} = \frac{-h_{fe}R_L}{h_{ie}+R_L(h_{ie}h_{oe}-h_{fe}h_{re})} = \frac{-h_{fe}}{h_{ie}G_L+(h_{ie}h_{oe}-h_{re}h_{fe})}$$

The gain of each stage may be calculated by breaking the circuit as follows:

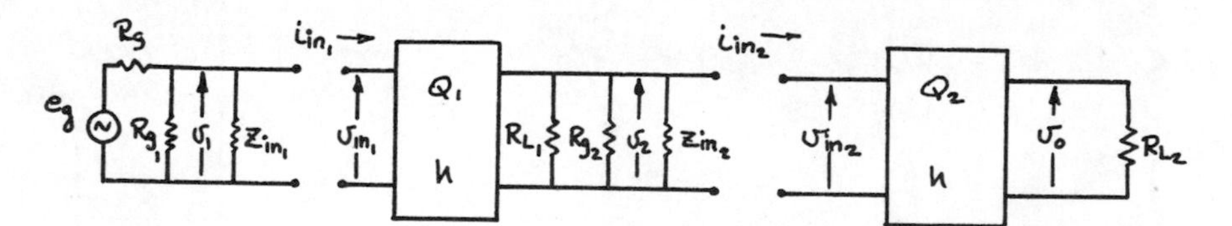

Definitions:

v_1 = Voltage at base of Q_1

$v_1 = v_{in_1}$

v_2 = Voltage at collector of Q_1

$v_2 = v_{in_2}$

v_o = Voltage across load R_{L_2}

$$v_1 = e_g \frac{(R_{g_1} \text{ in parallel with } Z_{in_1})}{R_{g_1} + (R_{g_1} \text{ in parallel with } Z_{in_1})}$$

Calculation of stage Q_2 (Eqn II):

$$\frac{v_o}{v_{in_2}} = A_v = \frac{-h_{fe}}{h_{ie}G_L + (h_{ie}h_{oe} - h_{oe}h_{fe})}$$

$$= \frac{-32}{(1670)(0.5\times10^{-4}) + (1670)(3.3\times10^{-5}) - (11.7\times10^{-4})(32)} = -320$$

To calculate gain of first stage, the impedance Z_{in_2} must first be obtained; this appears in parallel with R_{L_1} and R_{g_2}, all of which appear as a parallel load for stage Q_1. Then Z_{in_2} (from Eqn I) is given as:

$$Z_{in_2} = h_{ie} - \frac{h_{re}h_{fe}}{h_{oe}+G_L} = 1670 - 450 = 1220\ \Omega$$

Then the effective load for Q_1 is:

$$G'_{L_1} = \frac{1}{R'_{L_1}} = \frac{1}{R_{g_2}} + \frac{1}{R_{L_1}} + \frac{1}{Z_{in_2}} = \frac{1}{5K} + \frac{1}{10K} + \frac{1}{1.22K} = \frac{1}{895} = 1.12\times10^{-3}\ \mho$$

$$R'_{L_1} = 895\ \Omega$$

Then (from Eqn II):

$$\frac{V_2}{V_{in_1}} = A_v = \frac{-h_{fe}}{h_{ie}G'_L + (h_{ie}h_{oe} - h_{oe}h_{fe})}$$

$$= \frac{-32}{(1670)(1.12\times10^{-3}) + (1670)(3.3\times10^{-5}) - (11.7\times10^{-4})(32)} = -16.9$$

The total voltage gain from v_1 to v_o is:

$$\frac{v_o}{v_1} = \left(\frac{v_{in_2}}{v_1}\right)\left(\frac{v_o}{v_{in_2}}\right) = (-320)(-16.9) = 5400$$

To find the power delivered by the source, Z_{in_1} must be calculated (Eqn I):

$$Z_{in_1} = \frac{V_{in}}{i_{in_1}} = h_{ie} - \frac{h_{re}h_{fe}}{h_{oe}+G'_{L_1}}$$

$$= 1670 - \frac{(11.7\times10^{-4})(32)}{(3.3\times10^{-5}) + (1.12\times10^{-3})} = 1670 - 30 = 1640\ \Omega$$

Then the power delivered by the generator is:

$$P_{in} = \frac{V_1^2}{\text{Parallel combination of } R_{g_1} \text{ and } Z_{in_1}} = \frac{v_1^2}{\frac{(5\times10^3)(1640)}{5\times10^3 + 1640}} = \frac{v_1^2}{1.25\times10^3}$$

and the power delivered to the load is:

$$P_o = \frac{v_o^2}{R_{L_2}} = \frac{v_o^2}{20\times10^3}$$

why

Then the power gain is:

$$G = \frac{P_o}{P_{in}} = \frac{v_o^2/20\times10^3}{v_1^2/1.25\times10^3} = 0.0625\left(\frac{v_o}{v_1}\right)^2 = 0.0625(5400)^2$$

$$= 1.82\times10^6$$

$$G_{db} = 10 \log G = 62.6 \text{ db}$$

Now consider the inter-stage power loss (lost in parallel combination of R_{L_1} and R_{g_2}):

$$\frac{(R_{L_1})(R_{g_2})}{R_{L_1} + R_{g_2}} = \frac{(10\times10^3)(5\times10^3)}{10\times10^3 + 5\times10^3} = 3.33\times10^3\,\Omega$$

$$P_{I.L.} = \frac{v_2^2}{3.33\times10^3} = \frac{(16.9\,v_1)^2}{3.33\times10^3}$$

The total loss is then:

$$P_{Losses} = \frac{v_1^2}{5\times10^3} + \frac{(16.9\,v_1)^2}{3.33\times10^3}$$

But $v_1 = 0.556\,e_g$

$$P_{Losses} = \frac{(0.556 e_g)^2}{5\times10^3} + \frac{[(0.556)(16.9)e_g]^2}{3.33\times10^3}$$

$$= 27.7\times10^{-3} e_g^2$$

PROBLEM 2.

Subject: Electronics: Transistor circuit

Consider the following schematic diagram:

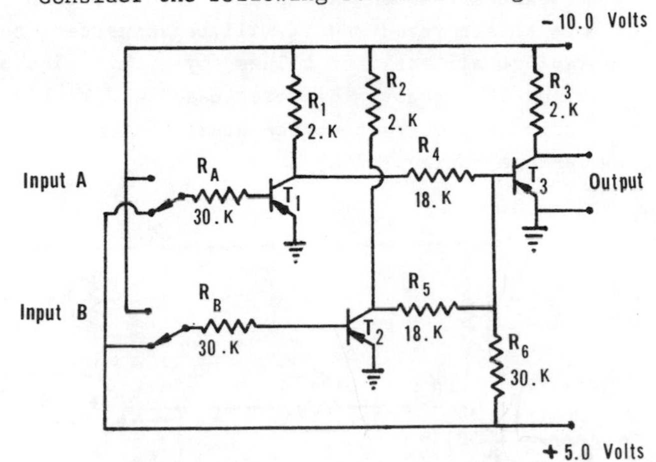

For the schematic diagram shown assume the following:

β (min) for transistors T_1 and T_2	20.
I_{co} (max) for any transistor	100. μa
V_{ce} (saturated) for any transistor (magnitude)	0.1 volt
V_{be} (max) for any "ON" transistor (magnitude)	0.35 volt
Resistor tolerance (max) from nominal values shown	± 10.%
Power Supplies	Constant Voltage

In the above schematic each input (A or B) may be switched to either the +5.0 volt or -10.0 volt supply.

For the WORST CASE conditions (all values at tolerance limit) determine the following two values:

(1) Wt. 7 What minimum β must transistor T_3 have for the output to be -0.1 volt when the inputs A and B are not alike?

(2) Wt. 3 What will be the minimum bias voltage on transistor T_3 when it is cut off?

(1) Assume switch "A" up and "B" down, then T_1 may be saturated and T_3 will be saturated; then replace all resistor values (by either plus or minus 10% to give the worst case) and T_2 by an equivalent current source equal to the cutoff leakage current of I_{co}:

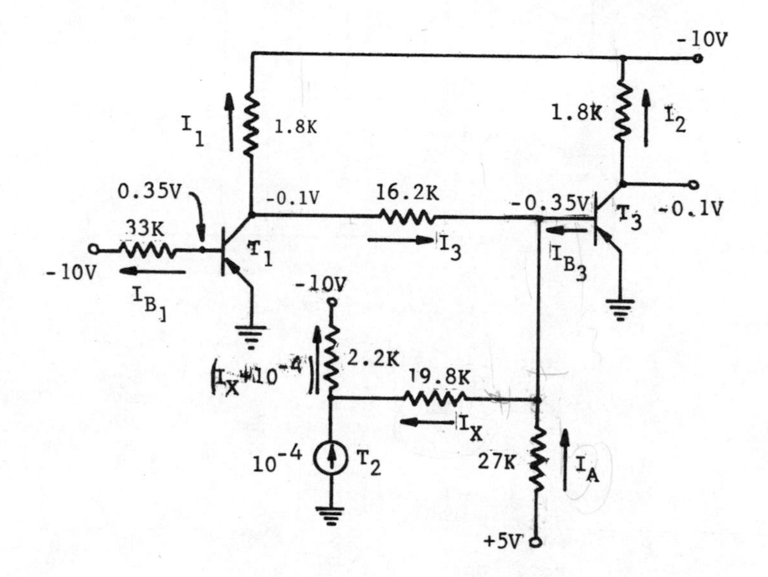

Then:

$$I_1 = \frac{10 - 0.1}{1.8 \times 10^3} = 5.5 \times 10^{-3} = I_2 \text{ (if saturated)}$$

$$I_{B_1} = \frac{10 - 0.35}{33 \times 10^3} = 0.292 \times 10^{-3}$$

β_1(minimum) = 20, so $I_{B_1}\beta_1 = (20)(0.29 \times 10^{-3})$
$= 5.84 \times 10^{-3}$

$\therefore I_{B_1} \beta_{1_{min}} > 5.5$, thus T_1 is indeed saturated.

$$I_A = \frac{5.35}{27 \times 10^3} = 0.198 \times 10^{-3}$$

Then I_X may be found from Kirchhoff's law. (around the voltage loop starting at the base of T_3):

$$+0.35 + 19.8 \times 10^3 I_X + 2.2 \times 10^3 (10^{-4} + I_X) - 10 = 0$$

$$I_X = 0.429 \times 10^{-3}$$

Then solving for T_3 base current:

$$I_{B_3} = -I_3 + I_X - I_A = -\frac{0.35 - 0.1}{16.2 \times 10^3} + 0.429 \times 10^{-3} - 0.198 \times 10^{-3}$$

$$= -\frac{(0.35 - 0.1)}{16.2 \times 10^3} + 0.429 \times 10^{-3} - 0.198 \times 10^{-3}$$

$$= 0.215 \times 10^{-3}$$

$$\therefore \beta_{3_{min}} = \frac{5.5 \times 10^{-3}}{0.215 \times 10^{-3}} = 25.6$$

$I_{B3} + I_A - I_X + I_3 = 0$

$I_{B3} = I_A + I_3 - I_A$

(2) Consider both switches to be "up", then:

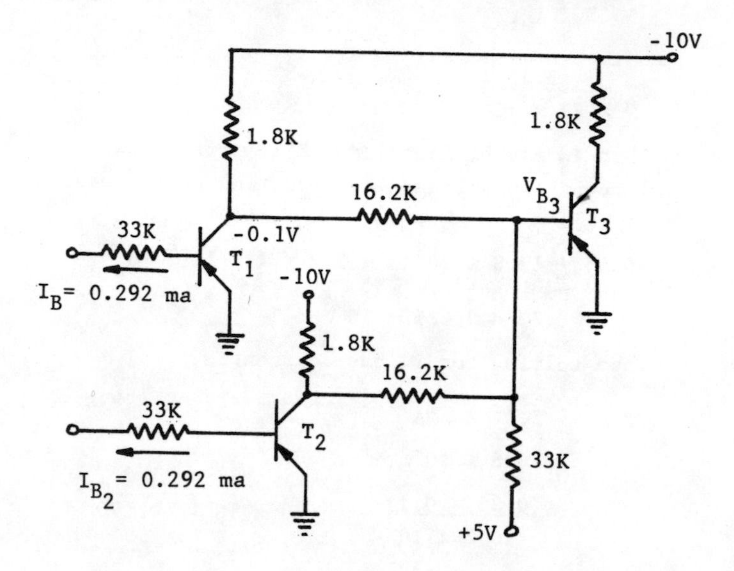

Then for V_{B_3}:

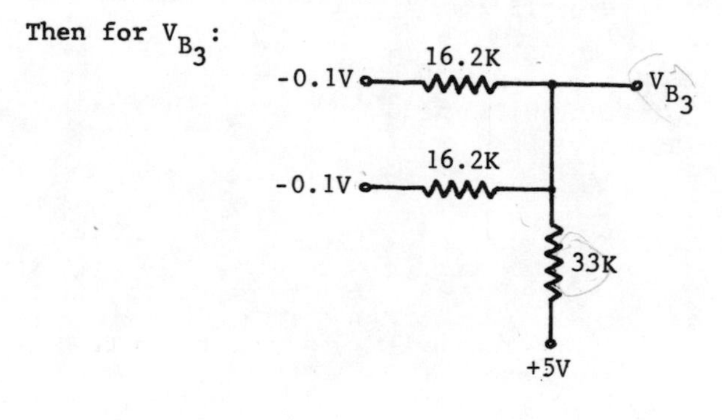

$$V_{B_3} = 5 - \frac{5.1 \times 33}{8.1 + 33} = +0.9 \text{ volts (min. bias)}$$

PROBLEM 3.

(1) Wt. 2 In an experimental triode vacuum tube, the cathode, grid, and anode are arranged in parallel planes with the grid half-way between the cathode and the anode. If the grid location is changed so that the grid is located nearer the cathode at a point that is 1/8 the distance from the cathode to the anode, the amplification factor will be altered by a factor of

a. - 1/4
b. - 1/2
c. - 1/ln4
d. - 1/ln2
e. - 1
f. - 2
g. - 4
h. - ln2
i. - ln4
j. - None of these

(2) Wt. 2 A resistance-capacity coupled amplifier has a coupling capacitor of 0.01 μf. If the size of the coupling capacitor is doubled, the result is to

a. - Lower the mid-frenquecy gain
b. - Raise the mid-frequency gain
c. - Raise the upper half-power frequency
d. - Lower the upper half-power frequency
e. - Double the lower half-power frequency
f. - Raise the lower half-power frequency by factor of 1.41
g. - Not alter lower half-power frequency
h. - Lower the lower half-power frequency by factor of .707
i. - Halve the lower half-power frequency
j. - None of these

(3) Wt. 2 A three phase, 60 cps rectifier uses three tubes supplied from a transformer bank connected with zig-zag secondaries. The lowest dominant ripple frequency in the output voltage will be a frequency, in cycles per second, of

a. - 90
b. - 120
c. - 150
d. - 180
e. - 210
f. - 240
g. - 270
h. - 300
i. - 330
j. - None of these

(4) Wt. 2 An experimental diode tube uses a thoriated-tungsten filament. Increasing the filament current by 20% will alter the operating conditions of the filament as follows:

a. - Increase temperature by 20%
b. - Increase temperature by 30%
c. - Increase temperature by 40%
d. - Increase temperature by 50%
e. - Increase temperature by 60%
f. - Increase emission by 30%
g. - Increase emission by 40%
h. - Increase emission by 50%
i. - Increase emission by 60%
j. - None of these

(5) Wt. 2 A transistor has the following parameters at 25°C when operated with a collector to base voltage of 5 volts and an emitter current of 1 milliampere in a base grounded configuration:

Input Impedance (output short-circuited)	55 ohms
Voltage Feedback Ratio (input short-circuited)	$2x10^{-4}$
Current amplification (output short-circuited)	0.980
Output admittance (input open-circuited)	0.2 μmho

When operated with a grounded emitter configuration with an emitter current of 2.0 ma., the base current is approximately

a. - 5.0 μa
b. - 10.0 μa
c. - 20.0 μa
d. - 30.0 μa
e. - 40.0 μa
f. - 50.0 μa
g. - 60.0 μa
h. - 80.0 μa
i. - 100.0 μa
j. - None of these

xxxxxxxxxxxxxxx

(1) By an analysis of the electrostatic field about a shielding screen of parallel wires, a voltage relationship may be obtained that will yield the desired results, i.e., the ratio of plate voltage to grid voltage. Consider the following diagram and potential relationships*:

* Spangenberg, "Vacuum Tubes", McGraw-Hill, Chap 7, Art. 7.2.

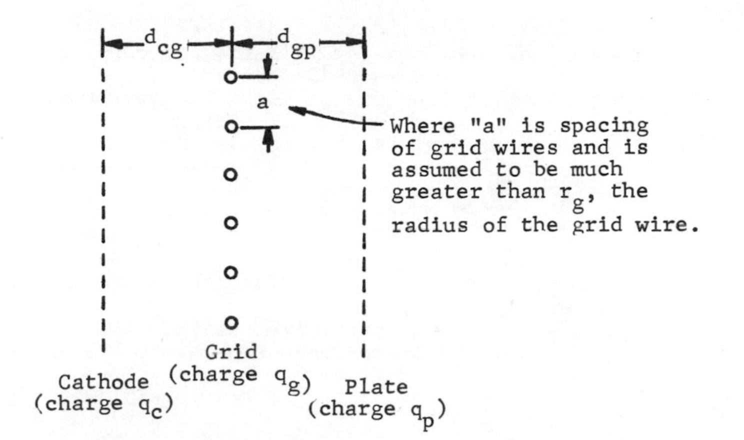

For a cathode reference potential of zero volts, the plate and grid potentials may be determined* in terms of the spacing and electrostatic charge on them as:

$$V_p = -\frac{d_{gp} q_g}{a\epsilon_o} - \frac{(d_{cg} + d_{gp})\, q_c}{a\epsilon_o}$$

$$V_g = -\frac{\ln(2\sin\frac{\pi r_g}{a})\, q_g}{2\pi\epsilon_o} - \frac{d_{cg} q_c}{a\epsilon_o}$$

And since the amplification factor "μ" of a tube is the ratio of plate voltage to the negative of grid voltage for a condition of cutoff (i.e., when cathode charge is zero), the above equations yield:

$$\mu = \frac{-2\pi d_{gp}}{a \ln(2\sin\frac{\pi r_g}{a})}$$

$$= K\, d_{gp} \quad \text{(for conditions of our problem)}$$

---

* Ibid, p 128.

Thus, for our problem let μ_1 correspond to the midway grid location (d_{gp_1}) and μ_2 correspond to the new location d_{gp_2}, 1/8 of the way between the cathode and plate, then:

$$\frac{\mu_1}{\mu_2} = \frac{(4/8)}{(7/8)} = \frac{4}{7}$$

$$\mu_2 = \frac{7}{4}\mu_1 = 1.75\mu_1$$

Thus none of the answers given satisfy the solution for the assumptions made.

However, using another approach to analysis* (i.e., assuming the grid to be a parallel plate with part of its area removed) and defining the amplification factor as the ratio of the coefficient of capacitance c_{21} to the coefficient of capacitance c_{31}, it can be shown (by making certain approximations) that for the corresponding figure given that:

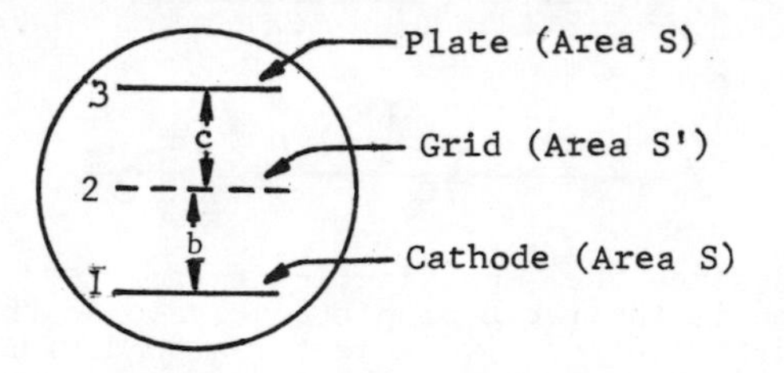

$$c_{21} = -\frac{\epsilon_v S'}{b}$$

$$c_{31} = -\frac{\epsilon_v (S-S')}{b+c}$$

then:

$$\mu = \frac{c_{21}}{c_{31}} = \left(\frac{S'}{S - S'}\right)\frac{(b+c)}{b}$$

---

* Booker, "An Approach to Electrical Science", McGraw-Hill, pp 149-151.

And for our problem statement: $\mu = K \frac{1}{b}$

then: $\frac{\mu_1}{\mu_2} = \frac{K/(4/8)}{K/(1/8)}, \quad \mu_2 = 4\mu_1$

which would satisfy answer g.

(2) Consider the equivalent circuit* of an R-C coupled triode amplifier: *also FET circuits get data sheet on FETs.*

And for the mid-frequency case:

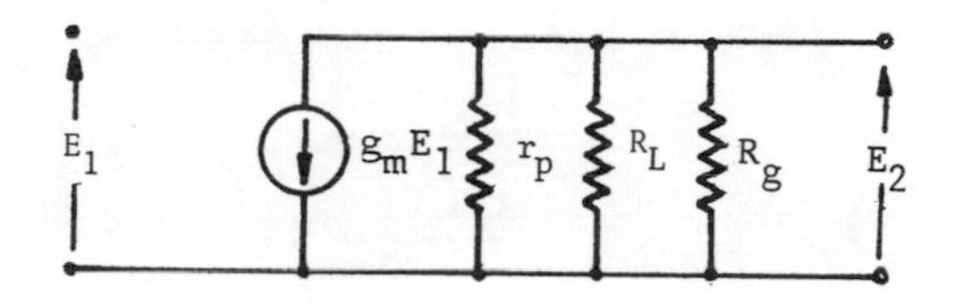

Define

$$\frac{1}{Req} = \frac{1}{r_p} + \frac{1}{R_L} + \frac{1}{R_g}$$

Then the mid-frequency gain can be found from:

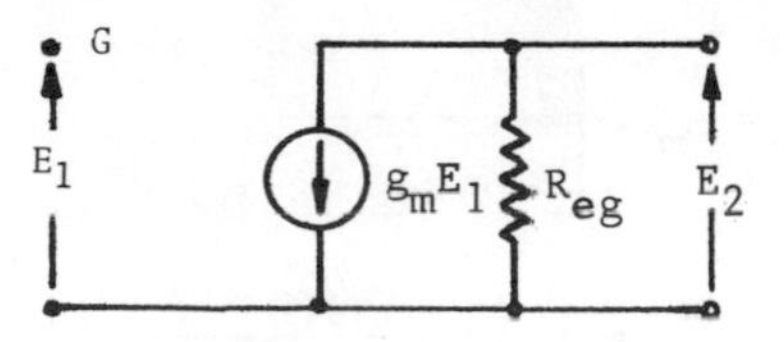

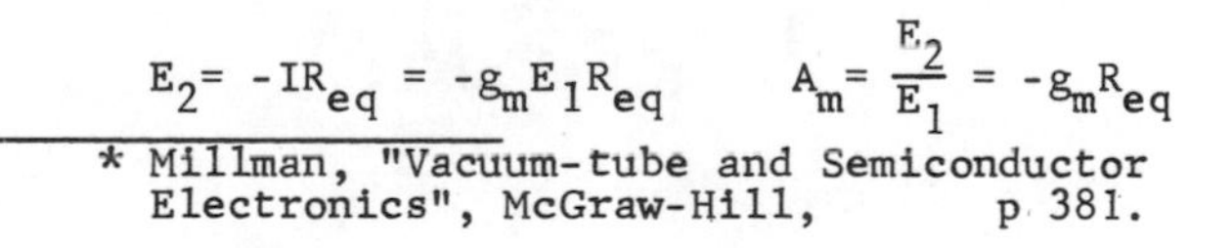

$$E_2 = -IR_{eq} = -g_m E_1 R_{eq} \qquad A_m = \frac{E_2}{E_1} = -g_m R_{eq}$$

* Millman, "Vacuum-tube and Semiconductor Electronics", McGraw-Hill, p. 381.

Now consider the low-frequency case:

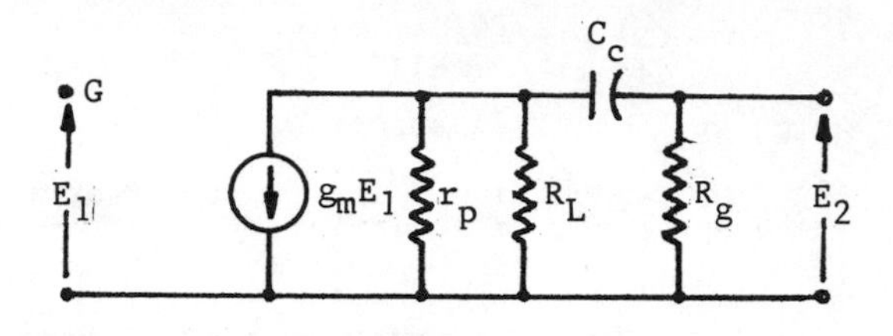

Which can be reduced to (by Thevenins Theorem):

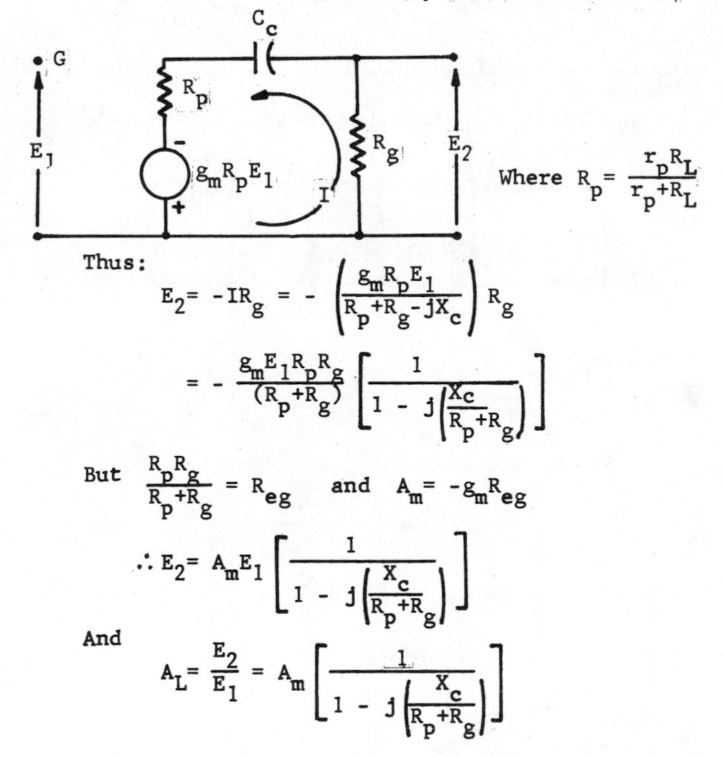

Where $R_p = \frac{r_p R_L}{r_p + R_L}$

Thus:

$$E_2 = -IR_g = -\left(\frac{g_m R_p E_1}{R_p + R_g - jX_c}\right) R_g$$

$$= -\frac{g_m E_1 R_p R_g}{(R_p + R_g)} \left[\frac{1}{1 - j\left(\frac{X_c}{R_p + R_g}\right)}\right]$$

But $\frac{R_p R_g}{R_p + R_g} = R_{eg}$ and $A_m = -g_m R_{eg}$

$$\therefore E_2 = A_m E_1 \left[\frac{1}{1 - j\left(\frac{X_c}{R_p + R_g}\right)}\right]$$

And

$$A_L = \frac{E_2}{E_1} = A_m \left[\frac{1}{1 - j\left(\frac{X_c}{R_p + R_g}\right)}\right]$$

Recall that the half-power point is where the voltage gain is reduced by a factor of $1/\sqrt{2}$:

$$A_L = \frac{1}{\sqrt{2}} A_m = A_m \left[\frac{1}{1 - j\left(\frac{X_c}{R_p + R_g}\right)} \right]$$

$$\therefore \frac{X_c}{R_p + R_g} = 1 = \frac{1}{2\pi f c_c (R_p + R_g)}$$

Thus if c_c is doubled, f is halved, or, answer "i" is correct.

(3) The lowest dominant ripple frequency in the output voltage will be a frequency, in cycles per second, of "d" - 180 -- which is three times the supply frequency of 60 cps.

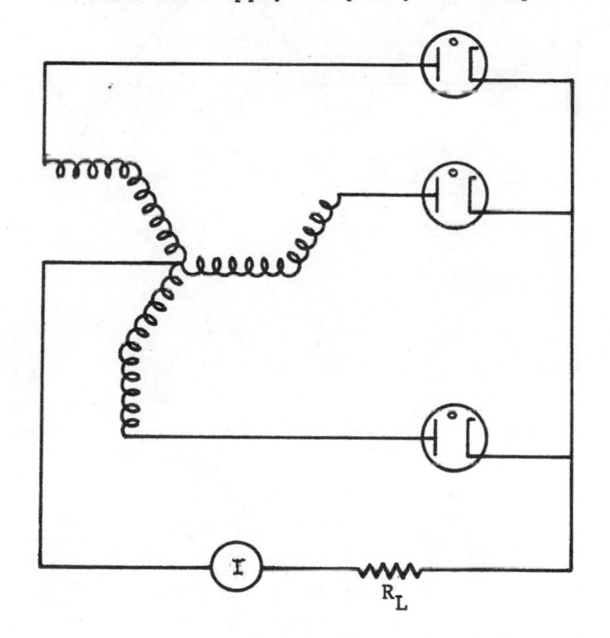

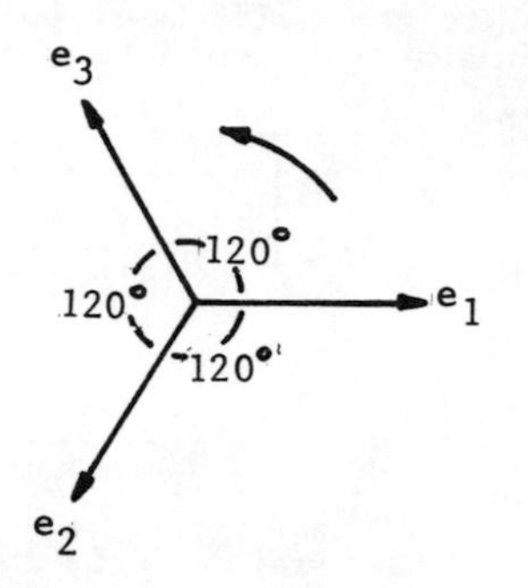

e_3
120°
120°
120°
e_1
e_2

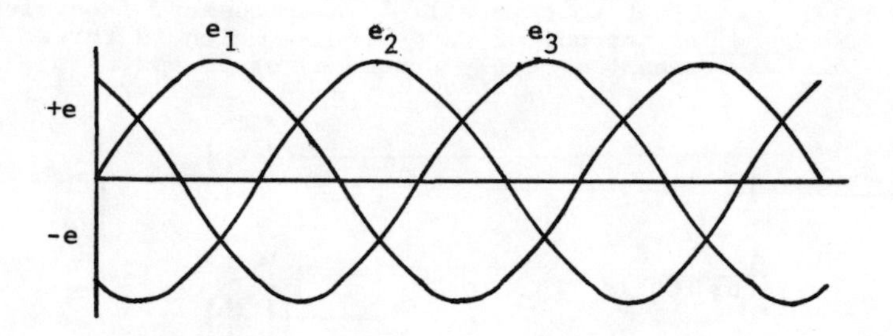

e_1
e_2
e_3
+e
-e

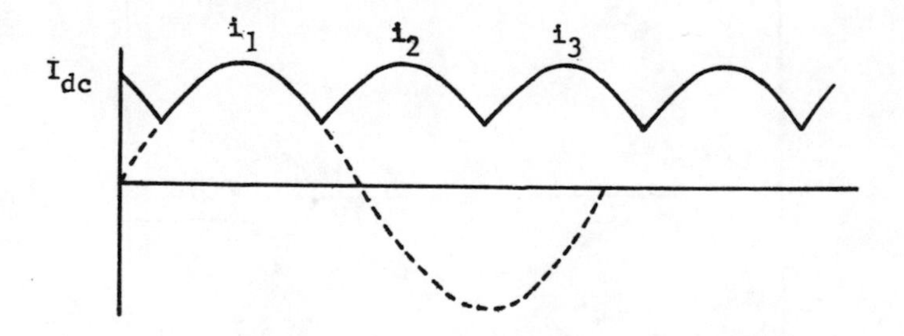

i_1
i_2
i_3
I_{dc}

(4) The emission current (assuming that the diode is not space charge limited) is given by Dushman's law as:

$$J = A_o T^2 \epsilon^{-b_o/T} \quad \text{amps}/\bar{m}^2$$

where J = emitted current density
T = temperature of source, Kelvin
A_o = proportionality constant (3×10^4 for thoriated-tungsten)
b_o = relationship to work function ($30{,}500^\circ$ for thoriated-tungsten)

The energy supplied to the filament will cause the temperature to rise until equilibrium is reached, i.e., the rate of heat loss is just equal to the rate of heat input. Since most of the heat loss is due to radiation*, the Stefan-Boltzman radiation law can be used and is given by:

$$W = e_T \sigma (T^4 - T_o^4) \cong e_T \sigma T^4 \quad \text{watts}$$

where T = Temp. of surface, Kelvin
T_o = Temp. of surrounding area, Kelvin
σ = S-B constant (5.77×10^{-8} W/$\bar{m}^2$/deg K^4)
e_T = emissivity of the surface
and $T^4 >> T_o^4$

Since the power radiated is approximately equal to the filament power input $(I_f^2 R_f)$:

$$T^4 \cong (KI_f)^2 \qquad T = \sqrt{KI_f}$$

And for an increase in filament current of 20%:

$$\frac{T_1}{T_2} = \sqrt{\frac{I_{f_1}}{I_{f_2}}} = \sqrt{\frac{I_{f_1}}{1.2 I_{f_1}}} = \sqrt{\frac{1}{1.2}}$$

$$\therefore T_2 = \sqrt{1.2}\, T_1$$

---

* Ryder, "Electronic Fundamentals And Applications", Prentice-Hall, p 77.

Thus answers a, b, c, d, and e will not be satisfied.

Consider the emission current density J:

$$J_1 = A_o T_1^2 \epsilon^{-b_o/T_1}$$

and for the 20% increase of filament current:

$$J_2 = A_o T_2^2 \epsilon^{-b_o/T_2} = A_o\left(\sqrt{1.2}\; T_1\right)^2 \epsilon^{-b_o/\sqrt{1.2}T_1}$$

$$\therefore \frac{J_2}{J_1} = 1.2\epsilon^{\frac{-b_o}{\sqrt{1.2}T_1}}\, \epsilon^{\frac{+b_o}{T_1}} = 1.2\epsilon^{-\frac{b_o}{T_1}(\frac{1}{\sqrt{1.2}} - 1)}$$

$$= 1.2\,\epsilon^{\frac{+b_o}{T_1}(0.085)}$$

For normal operating temperatures of T_1 (say 2,000° K)

$$\frac{J_2}{J_1} = 1.2\,\epsilon^{\frac{30,500}{2,000}(0.085)} = 1.2\text{x}3.65 = 4.38$$

This is well over 400% of the original emission current, thus none of the answers given are correct. In general, at normal operating temperatures, the exponential term in the emission equation accounts for most of the variation of emission with temperature*.

---

* Spangenberg, "Vacuum Tubes", McGraw-Hill, p 31.

(5) The given parameters merely define the well-known "h" parameters* in the common base configuration:

$$Z_{in} \text{ defines } h_{11_b} = \left.\frac{v_1}{i_1}\right|_{v_2=0} = 55\,\Omega$$

Voltage feedback ratio defines

$$h_{12} = \left.\frac{v_1}{v_2}\right|_{i_1=0} = 2 \times 10^{-4}$$

(Note: this is defined as the fraction of the output voltage at the input <u>with the input open circuited</u>; thus the problem statement is in error, since for a short circuit input the ratio would obviously be zero.)

Current amplification defines

$$h_{21} = \left.\frac{i_2}{i_1}\right|_{v_2=0} = -0.980$$

Output admittance defines

$$h_{22} = \left.\frac{i_2}{v_2}\right|_{i_1=0} = 0.2\,\mu\mho$$

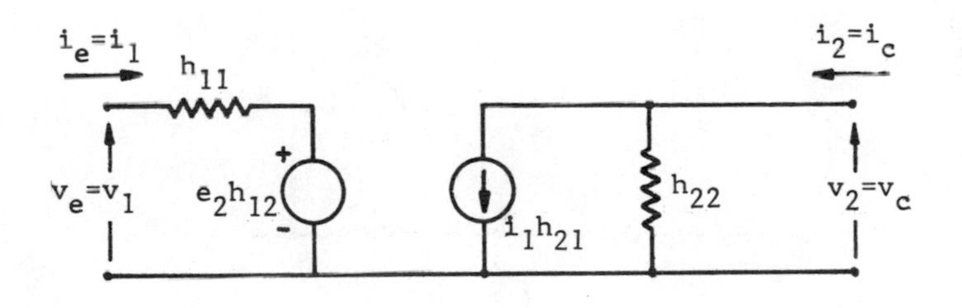

* Millman, "Vacuum-tube and Semiconductor Electronics", McGraw-Hill, p 255.

Or, in terms of the more familiar "T" equivalent:

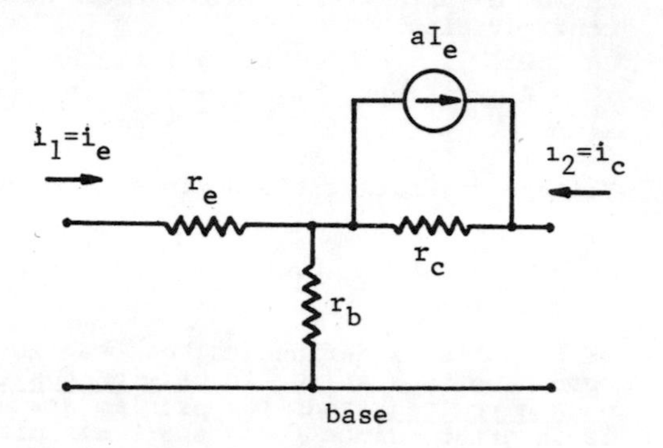

where: $a = -h_{21} = 0.98$

$$r_c = 1/h_{22} = 5 \times 10^6$$

$$r_b = \frac{h_{12}}{h_{22}} = 10^3$$

$$r_e = h_{11} - \frac{h_{12}(1 + h_{21})}{h_{22}} = 35$$

Now consider the common emitter configuration:

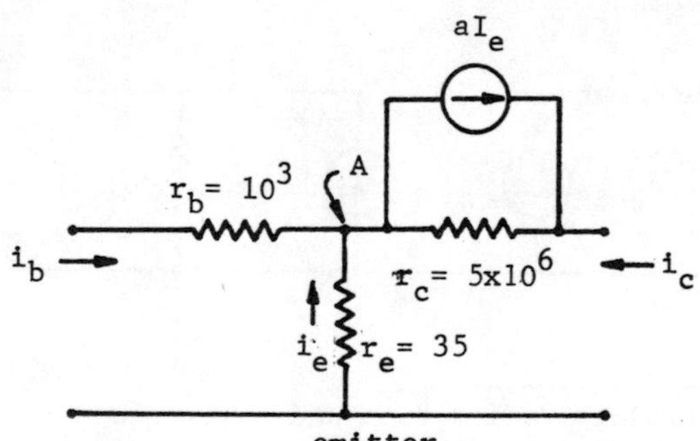

To solve for the value of i_b, the sum of the currents at junction A are:

$$i_b + i_c + i_e = 0$$

and since i_e is given as 2 ma, the voltage drop across r_e is only $(2 \times 10^{-3})(35) = 0.07$ v and assuming the voltage between collector and the common base is 5 volts, the current through $r_c \cong \frac{5 \text{ volts}}{5 \times 10^6} = 1\ \mu a$, thus $i_c \cong -aI_e$, then:

$$i_b = -(i_c + i_e) \cong -(-aI_e + I_e)$$

$$\cong -(1 - a)I_e = -(1 - 0.98)2 \times 10^{-3} \text{ amps}$$

$$\cong -0.02 \times 2 \times 10^{-3} = -0.04 \times 10^{-3} = -40\ \mu A$$

Thus "e" is the approximate solution.

PROBLEM 4.

Subject: Communications: Telephone circuit

In the figure below, the simple low pass filter network consisting of the balanced inductance (L) and the capacitor (C) is used to provide isolation between a telephone set bridged on the wire line and the carrier frequency equipment operating on the same wire line. The designed cut-off frequency of this low pass filter is 3.0 kcs. Determine and sketch the impedance as seen by the wire line and looking from the wire line into the filter network as a function of frequency over the voice range for a telephone set having an impedance of 600 ohms when connected or in use under the conditions:

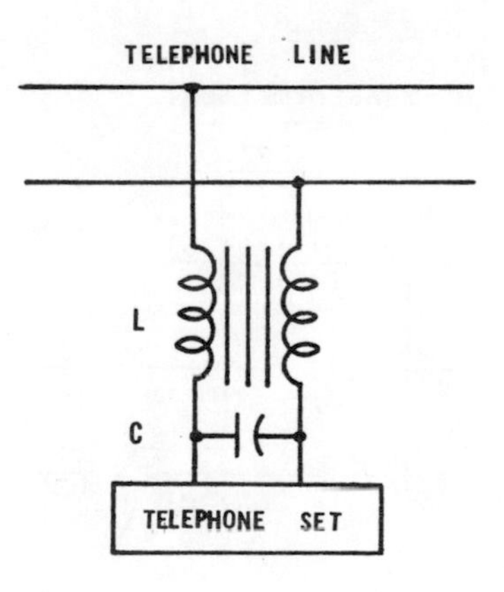

(1) <u>Wt. 5</u> With the telephone set connected or in use.

(2) <u>Wt. 5</u> With the telephone set disconnected or not in use.

The question arises, are the two inductances to have coupling between them or not? Since the statement is made, a "simple low pass filter ..", any mutual inductance between the L's is not simple.

Proceeding on the basis of M = 0

Balanced circuit:

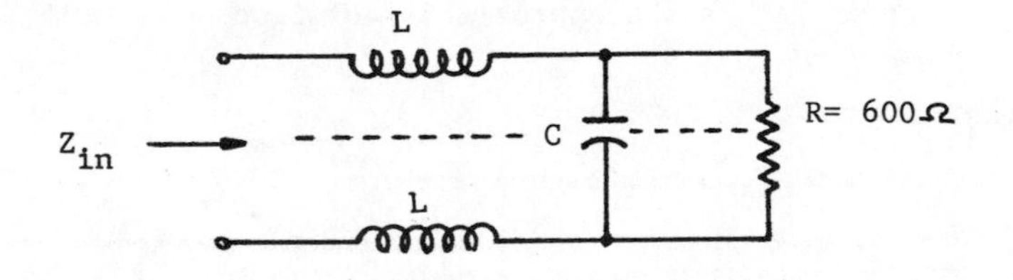

Unbalanced equivalent:

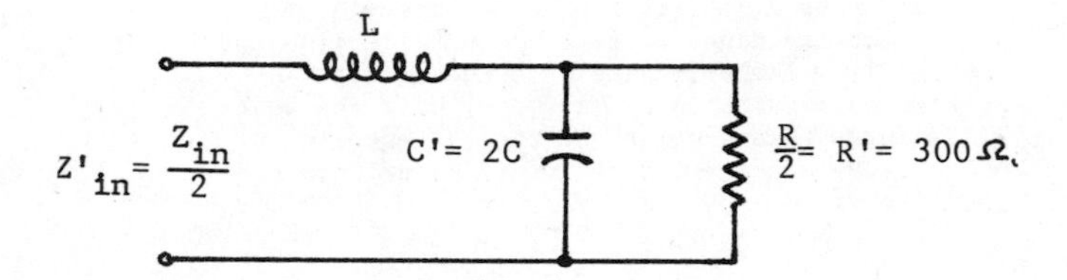

By Laplace Transformation:

$$Z'_{in} = sL + \frac{R'}{1 + sR'C'}$$

$$= \frac{(s + a)(s + b)}{(s + c)}$$

where $a = \frac{1}{2R'C'} - \frac{1}{4R'^2C'^2} - \frac{1}{LC'}$

$b = \frac{1}{2R'C'} + \frac{1}{4R'^2C'^2} - \frac{1}{LC'}$

$c = \frac{1}{R'C}$ note $c > b > a$

we assume a and b are real and $R^2 < \frac{L}{4C'}$

Using the method of asymptotes, Z'_{in} may be plotted for $s(j\omega)$ very, very low

$Z'_{in} = L\frac{as}{c} = R'$ $\omega = 2\pi f$ $j = \sqrt{-1}$

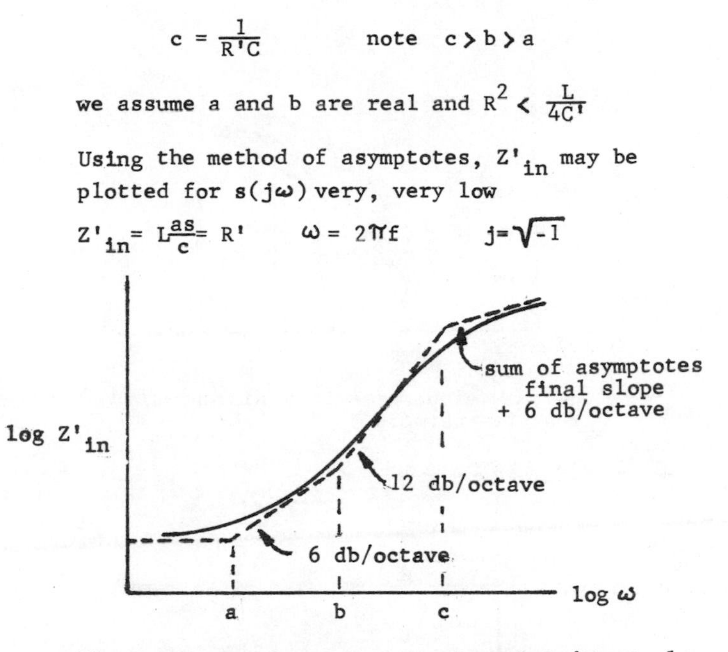

First corner occurs at s = a Asymptote has + slope 6 db/octave.
Next corner occurs at s = b Asymptote has + slope
Third corner occurs at s = c Asymptote has - slope
The solid line is then log Z'_{in} plotted vs, log ω
The departure of Z'_{in} at "corners" is 3 db.
Since cut-off frequency is stated as 3 KC,
"a" has a value of $2\pi \times 3 \times 10^3 = 6\pi \times 10^3$ rad/sec.
Now $Z'_{in} = 1/2\ Z_{in}$; also by proper choice of L and C the corners a and b may be moved closer together.

Suppose $R^2 = \frac{L}{4C}$

then $a = b = \frac{c}{2}$ and $Z'_{in} \equiv \frac{(s + a)^2}{s + c}$

In this case the first "corner" occurs at $a = \omega$ and the asymptotic slope between a and c is 12 db/octave.

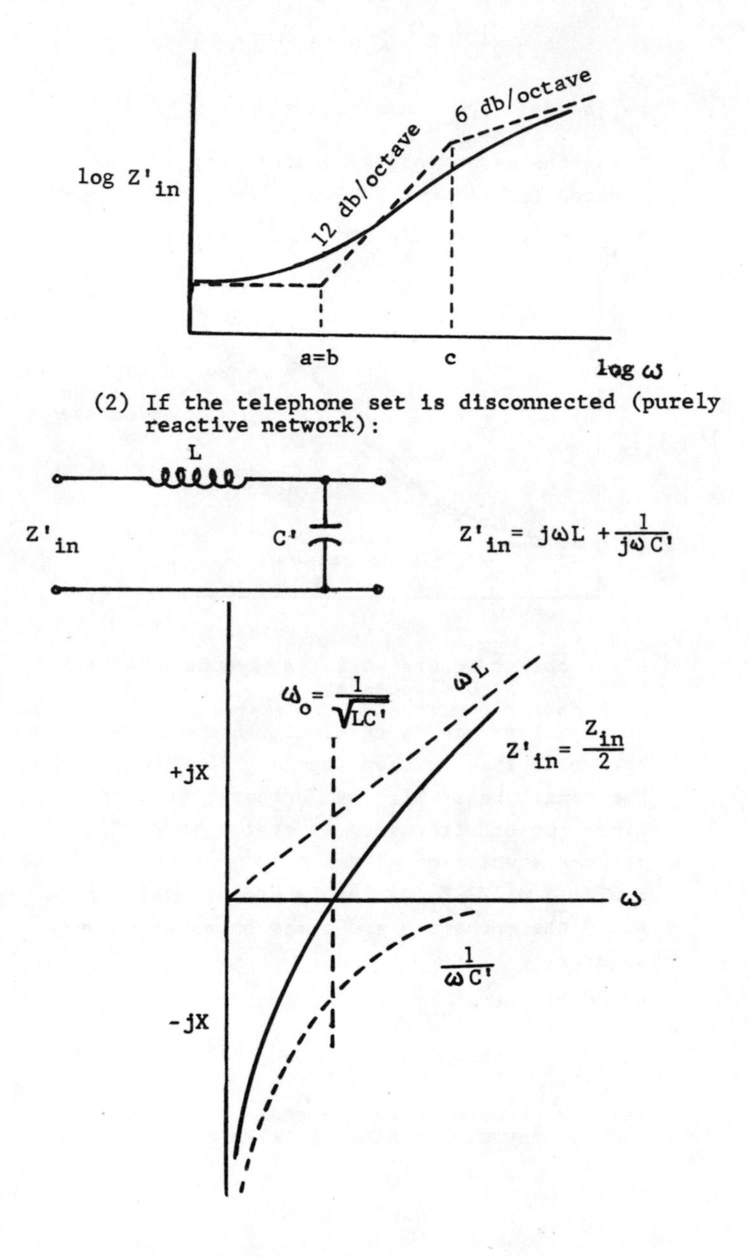

(2) If the telephone set is disconnected (purely reactive network):

$$Z'_{in} = j\omega L + \frac{1}{j\omega C'}$$

PROBLEM 5.

In the modulating circuit sketched below, the modulating signal (see spectrum sketch) is limited to angular frequencies ω where

$$\omega_{m_1} < \omega < \omega_{m_2} \lll \omega_c.$$

Where ω_m = modulating frequency; and ω_c = carrier frequency.

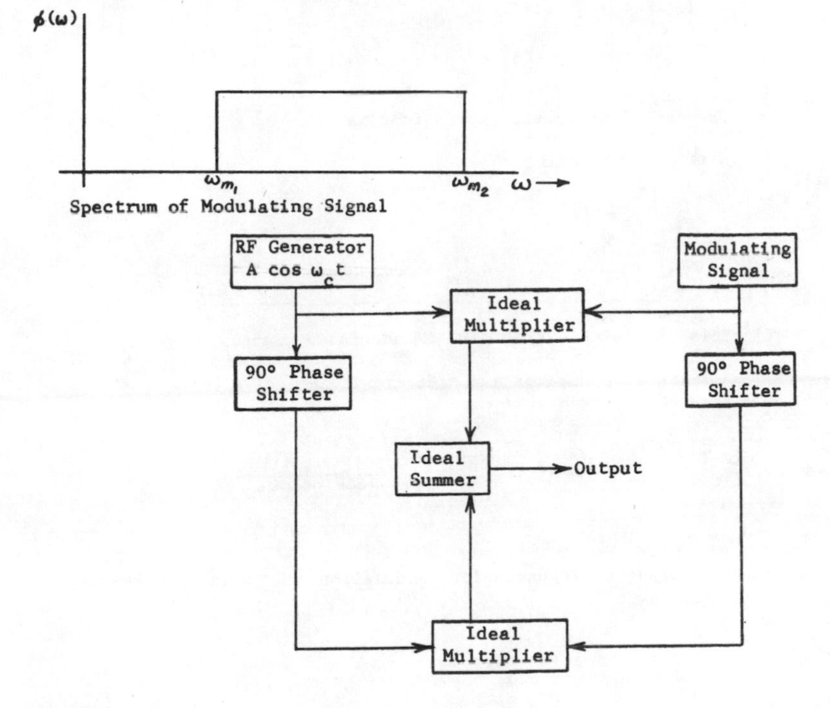

Spectrum of Modulating Signal

REQUIRED:

Wt.		
5	(a)	Sketch the spectrum of the output.
2	(b)	What is this type of modulated signal called?
3	(c)	Sketch another circuit which would produce the same output spectrum.

(a) The in-phase input is as follows:

$A \cos \omega_c t \cdot \phi \cos \omega_m t =$

$\frac{A\phi}{2} \left[\cos (\omega_c - \omega_m)t - \cos (\omega_c + \omega_m)t\right]$

The 90° phase shift is as follows:

$A \sin \omega_c t \cdot \phi \sin \omega_m t =$

$\frac{A\phi}{2} \left[\cos (\omega_c - \omega_m)t + \cos (\omega_c + \omega_m)t\right]$

The sum of above two expressions is:

$A \phi \cos (\omega_c - \omega_m)t$

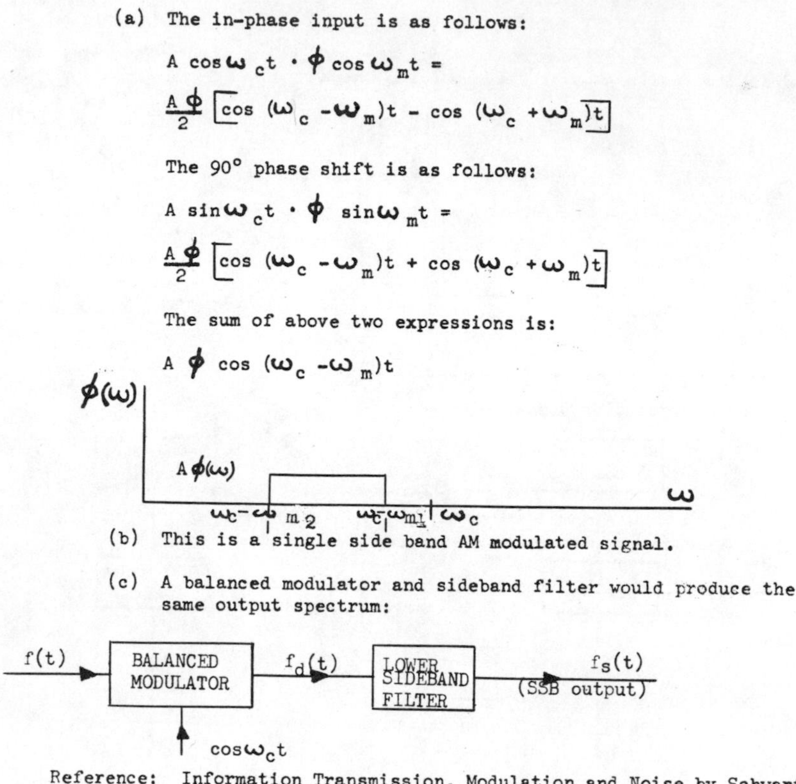

(b) This is a single side band AM modulated signal.

(c) A balanced modulator and sideband filter would produce the same output spectrum:

f(t) → BALANCED MODULATOR → $f_d(t)$ → LOWER SIDEBAND FILTER → $f_s(t)$ (SSB output)

$\cos \omega_c t$

Reference: Information Transmission, Modulation and Noise by Schwartz, McGraw-Hill p. 106, 107.

PROBLEM 6.

A certain transmitter has an effective radiated power of 9 KW with the carrier unmodulated and 10.125 KW when the carrier is modulated by a sinusoidal signal.

REQUIRED:

Wt.		
5	(a)	Determine the percent modulation at 10.125 KW output.
5	(b)	Determine the total effective radiated power if in addition to the sinusoidal signal, the carrier is simultaneously modulated 40% by an audio wave.

(a) Assume transmission is through a 1 ohm resistance. For the unmodulated carrier with carrier frequency ω_o:

$$v_o(t) = A \cos\omega_o t$$

$$P = \int v_o(t) = \frac{A^2}{2} = 9 \text{ KW, since:}$$

$$P = \frac{1}{2\pi}\int_0^{2} \frac{v^2(t)}{R}\,dt = \frac{A^2}{2\pi}\int_0^{2\pi} \cos^2\omega_o t dt$$

$$= \frac{A^2}{2\pi}\int_0^{2} \frac{\cos 2\omega_o t + 1}{2}\,dt = \frac{A^2}{2\pi}\cdot\frac{2\pi}{2} = \frac{A^2}{2}$$

Then, for the modulated carrier with modulating frequency ω_1, and modulating coefficient "m", we obtain

$$v_o(t) = A\,(1 + m \cos\omega_1\, t)\cos\omega_o t$$

$$= A \cos\omega_o t + \frac{Am}{2}\cos(\omega_o + \omega_1)t + \frac{Am}{2}\cos(\omega_o - \omega_1)t$$

since a basic trigonometric relation states:

$$\cos\alpha\,\cos\beta = \frac{1}{2}\cos(\alpha + \beta) + \frac{1}{2}\cos(\alpha - \beta)$$

Therefore: $P = \dfrac{A^2 + \left(\frac{Am}{2}\right)^2 + \left(\frac{Am}{2}\right)^2}{2} = 9\left[(1 + 2\,\frac{m}{2})^2\right] = 10.125$

$\frac{m}{2} = \sqrt{\frac{1.125}{18}} = 0.25$ and $m = 0.5$, or 50% ANS.

(b) Denoting the audio wave frequency with ω_2, we obtain:

$$v_o(t) = A\,(1 + 0.5 \cos\omega_1\, t + 0.4 \cos\omega_2 t)\cos\omega_o t$$

$$= A\cos\omega_o t + \frac{0.5A}{2}\cos(\omega_o + \omega_1)t + \frac{0.5A}{2}\cos(\omega_o - \omega_1)t$$

$$+ \frac{0.4A}{2}\cos(\omega_o + \omega_2)t + \frac{0.4A}{2}\cos(\omega_o - \omega_2)t$$

Then the effective radiated power is:

$$P = \frac{A^2 + 2\left(\frac{0.5A}{2}\right)^2 + \left(\frac{0.4A}{2}\right)^2}{2} = 10.125 + 9\left[2\left(\frac{0.4}{2}\right)^2\right]$$

$$= 10.125 + 0.72 = 10.845 \text{ KW}$$ ANS.

The frequency spectrum is as follows:

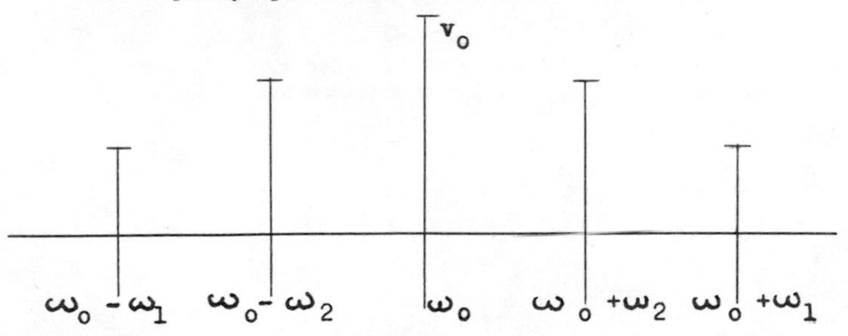

PROBLEM 7.

Information Theory

In order to specify the communications link for a closed circuit television system, the bit rate must be known.

The monochrome television picture signal of this system requires 10 distinct levels of brightness for good resolution. This television system also includes the following parameters:

(1) Frame rate, 15 frames per second
(2) Lines per frame, 1200
(3) Discrete picture elements, 100 per line

REQUIRED:

Determine the channel capacity in bits per second required to transmit the above video signal with all levels equally probable and with all elements assumed to vary independently.

List any assumptions that you make.

The number of different possible pictures is

$$P = 10 \times 10 \ldots \times 10 = 10^{1200 \times 100} = 10^{1.2 \times 10^5}$$

Probability of each element or picture is

$$= \frac{1}{10^{1.2 \times 10^5}}$$

The Channel Capacity is defined as

$$C = \lim_{T \to \infty} \frac{1}{T} \log_2 M(T)$$ where M(T) is the total number of messages in T seconds.

$$= S \log_2 P \quad \text{bit/sec}$$ where S is the signalling speed.

$S = 15$

$$C = 15 \log_2 10^{1.2 \times 10^5}$$

$$= 1.8 \times 10^6 \log_2 10$$

$$= 6.0 \times 10^6 \text{ bit/second}$$

Assumption: It is assumed that the signal to noise ratio is large or the error probability is small so the channel is deemed noiseless.

PROBLEM 8.

The circuit shown below is an N-channel Junction Field-Effect Transistor with self-bias and a pinch-off voltage of -3 volts. At that value of pinch-off voltage, the current is 6 mA. The breakdown voltage for this transistor is 30 volts.

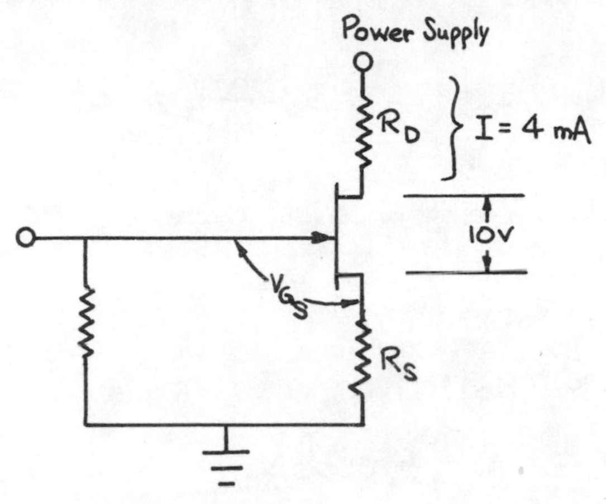

REQUIRED: Design the above circuit so that the device will be biased at approximately 10 V drain-to-source and have a channel current of approximately 4 mA.

Solution

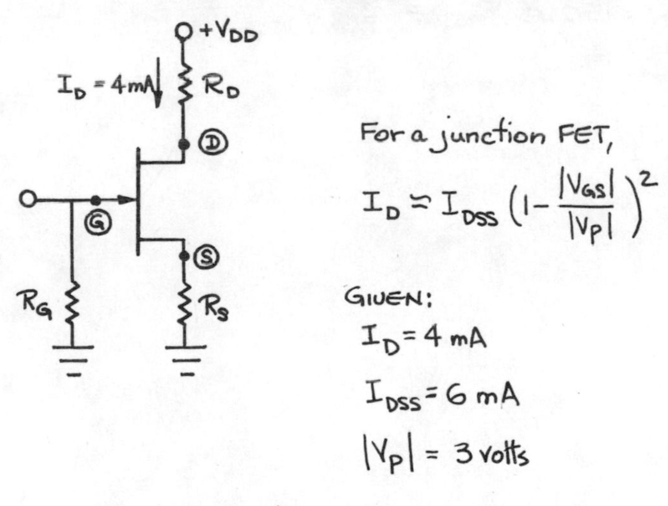

$$\therefore\ 4 \simeq 6\left(1-\frac{|V_{GS}|}{3}\right)^2 \Rightarrow |V_{GS}| \simeq 0.55 \text{ volts}$$

Gate leakage current I_{GSS} is typically of the order of nanoamperes. Choose $R_G = 1\,M\Omega$, so that the gate remains within millivolts of ground potential. Choose R_S to obtain the proper V_{GS}.

$$|V_{GS}| = 0.55 \text{ volts},\ I_D = 4\,mA \Rightarrow R_S = \frac{0.55}{4\times10^{-3}} \simeq 140\,\Omega$$

Since the breakdown voltage is 30 volts, choose $V_{DD} = 24$ volts.

$$V_{DS} = V_{DD} - I_D(R_D + R_S)$$

$$10 = 24 - 4\times10^{-3}(R_D + R_S)$$

$$R_D + R_S = \frac{14}{4\times10^{-3}} = 3500\,\Omega$$

$$\therefore R_D = 3500 - 140 = 3360\,\Omega$$

If 10% resistors are used, the available values are:

$$R_S = 150\,\Omega \text{ and } R_D = 3.3\,K\Omega$$

To [illegible] the resultant gain

7

CONTROL SYSTEMS

CONTROL SYSTEMS 1.

The open loop transfer function for a control system is approximated by:

$$G(s) = \frac{C(s)}{E(s)} = K \frac{s-3}{(s + 0.5)(s + 7)}$$

It is desired to make the output signal (C) correspond as nearly as possible to some input signal, (R), in steady state, at the same time keeping the system stable.

REQUIRED:

Wt.		
3	(a)	Sketch a block diagram for a feedback control system to accomplish the given objective. Carefully label the summation polarity of all signals coming into the feedback junction summing point. (Note that the given transfer function has peculiar properties.)
7	(b)	Select a value of K which assures system stability and at the same time brings the ratio $\frac{C}{R}$ in steady-state as close to + 1.0 as possible. (Note that the properties of G(s) are such that it is advisable to make a very careful check on the requirements for closed-loop system stability.)

a) The block diagram could be given as follows (with either positive or negative feedback):

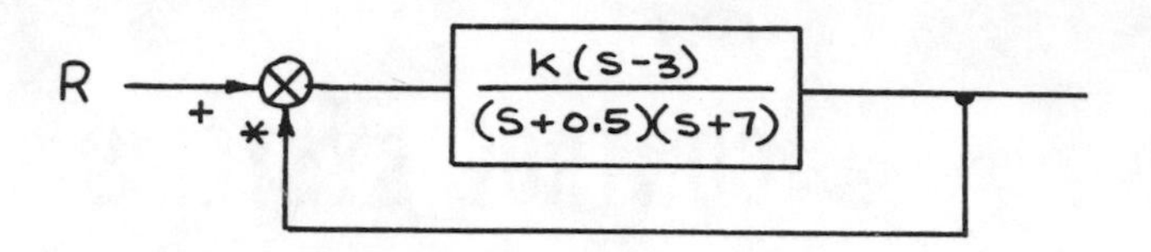

*For negative (−) feedback, the system function is:

$$G_{SYST} = \frac{G}{1+G} = \frac{K(S-3)}{(S+0.5)(S+7)+K(S-3)}$$

$$= \frac{K(S-3)}{S^2+(7.5+K)S+(3.5-3K)}$$

The denominator (which determines the "character" of the response) must not have any negative factors (indicating closed loop poles in the right half plane -- or an unstable system). A simple test for stability is the Routh criterion; or in this case, (for a simple second-order system) all of the coefficients of the denominator polynomial must be positive, therefore:

$$3.5-3K>0, \quad \therefore K<\frac{3.5}{3}=1.17$$

*For positive (+) feedback, the system function is:

$$G_{SYST} = \frac{G}{1-G} = \frac{K(S-3)}{(S+0.5)(S+7)-K(S-3)}$$

$$= \frac{K(S-3)}{S^2+(7.5-K)S+(3.5+3K)}$$

Here, for stability, again the denominator polynomial must be positive, thus:

$$K < 7.5$$

Of course the root locus method of analysis may also be used:

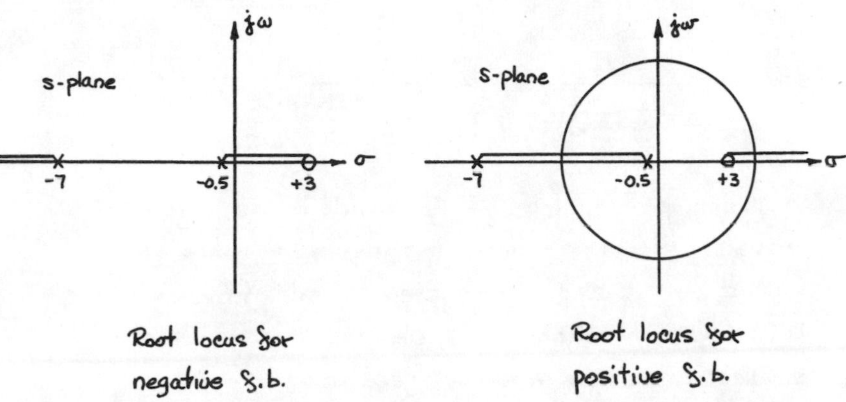

Root locus for negative f.b. (angles sum to ±180°)

Root locus for positive f.b. (angles sum to 0°)

b) For $\frac{C}{R}$ to be as close to unity as possible, then it is only necessary to minimize $E = R - C$ (since it is a unity feedback system):

For negative feedback:

$$E = \frac{R}{1 + G}$$

Now assume a step input $\left(R = \frac{r_o}{s}\right)$ and that one is interested in the error after a long period of time -- such that the final value theorem may be applied:

$$e\Big|_{t\to\infty} = \lim_{s\to 0} sE = \lim_{s\to 0} s\left[\frac{r_o/s}{1+\dfrac{K(s-3)}{(s+0.5)(s+7)}}\right]$$

$$= \frac{r_o}{1+\dfrac{K(-3)}{(0.5)(7)}} = \frac{r_o}{1-0.85K}$$

To minimize this, $(1-0.85K)$ should be as large as possible. However, since the maximum value of K that can be used (for system stability) makes $0.85K = 1$, therefore it is obvious that for the smallest error for negative feedback, $K \to 0$.

Consider positive feedback:

$$E = \frac{R}{1-G}$$

then

$$e\Big|_{t\to\infty} = \lim_{s\to 0} s\left[\frac{r_o/s}{1-\dfrac{K(s-3)}{(s+0.5)(s+7)}}\right]$$

$$= \frac{r_o}{1+\dfrac{3K}{3.5}}$$

and here K can take on values up to 7.5 for a

stable system, therefore the minimum value for "e" will be when K just equals 7.5:

$$e = \frac{r_o}{1+6.43} = \frac{r_o}{7.43}$$

Thus, for this peculiar configuration, positive feed back gives the smallest system error and the value of K may vary from 0 to 7.5.

CONTROL SYSTEMS 2.

Subject: Control: System stability

Consider the following simplified diagram for a control system:

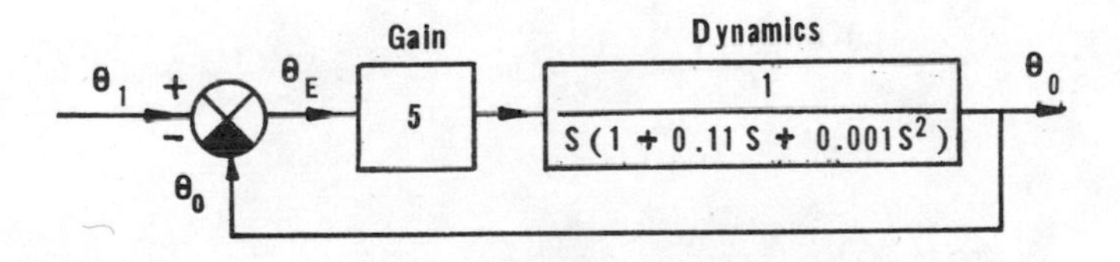

(1) Wt. 6 Determine whether the system is stable by any standard criterion.

(2) Wt. 4 What range of real values of gain will enable the system to be stable?

There are several methods of determining system stability, two of which are:

1. Routh's criterion*
2. Root locus**

All of these methods are based upon solving the equivalent closed loop "characteristic equation" for positive roots, i.e., all solutions that would indicate whether the resulting exponential terms were expanding with time.

Routh Criterion: Solve for the closed loop transfer function:

$$KG_{CL} = \frac{KG}{1+KG} = \frac{5}{0.001S^3 + 0.11S^2 + S + 5}$$

and the characteristic equation (denominator) can be formed into a triangular array as follows:

Characteristic eqn.:

$$A_n S^n + A_{n-1} S^{n-1} + A_{n-2} S^{n-2} + \text{----} \; A_o$$

Then:

S^n	A_n	A_{n-2}	A_{n-4}	- - -
S^{n-1}	A_{n-1}	A_{n-3}	A_{n-5}	- - -
S^{n-2}	X_a	X_b	X_c	- - -
⋮	⋮			
S^o	Z_a			

Where $X_a = \dfrac{A_{n-1}A_{n-2} - A_n A_{n-3}}{A_{n-1}}$ etc.

All coefficients of the first coefficient column must be of the same sign for stability.

---

* Thaler and Brown, "Servo-mechanism Analysis", McGraw-Hill, p 151.

** Ibid, pp 304 - 313.

Thus:

s^3	0.001	1
s^2	0.11	5
s^1	0.955	
s^o	5	

where

$$X_1 = \frac{(.11)(1) - (0.001)(5)}{0.11} = 0.955$$

$$Y_1 = \frac{(0.955)(5) - 0}{.955} = 5$$

Therefore system is stable for the given value of gain.

(2) To determine the range of values of gain for the system to be stable, merely let the "5" in the Routh-Hurwitz array be a variable K and solve:

s^3	0.001	1
s^2	0.11	K
s^1	X_1	

$$X_1 = \frac{(.11)(1) - (0.001)\,K}{0.11}$$

and since the requirement for X_1 is that it be positive for stability:

$$0.11 > 0.001\ K$$

$$K < 110 \text{ for stability}$$

Solution based upon root-locus method:

$$KG = \frac{K}{S(1 + 0.11S + 0.001S^2)} = \frac{K/1000}{S(1000 + 110S + S^2)}$$

$$= \left(\frac{K}{1000}\right)\left[\frac{1}{S(S + 10)(S + 100)}\right]$$

The root-locus plot will then be:

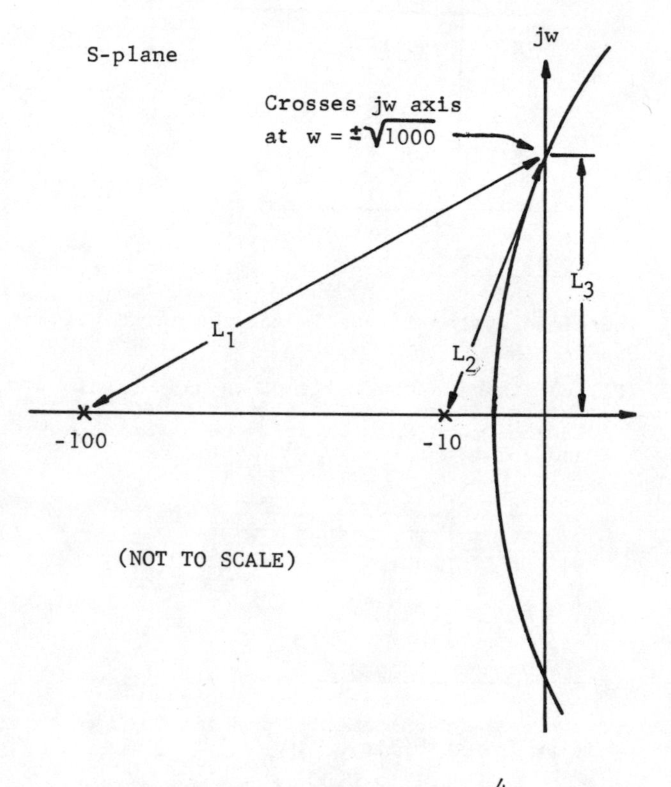

$$K_o = (L_1)(L_2)(L_3) = 11.0 \times 10^4$$

$$K = \frac{K_o}{1000} = 110$$

Thus for any value of K_o less than 110, the system would be stable.

CONTROL SYSTEMS 3.

In a constant temperature bath, a bridge circuit is employed as the error detector, as shown in the figure.

E is the temperature sensing element

R is a fixed resistor of 100 Ω

R_A and R_B are fixed resistors, the values of which are to be determined

P is a potentiometer of 500 Ω for setting the bath temperature

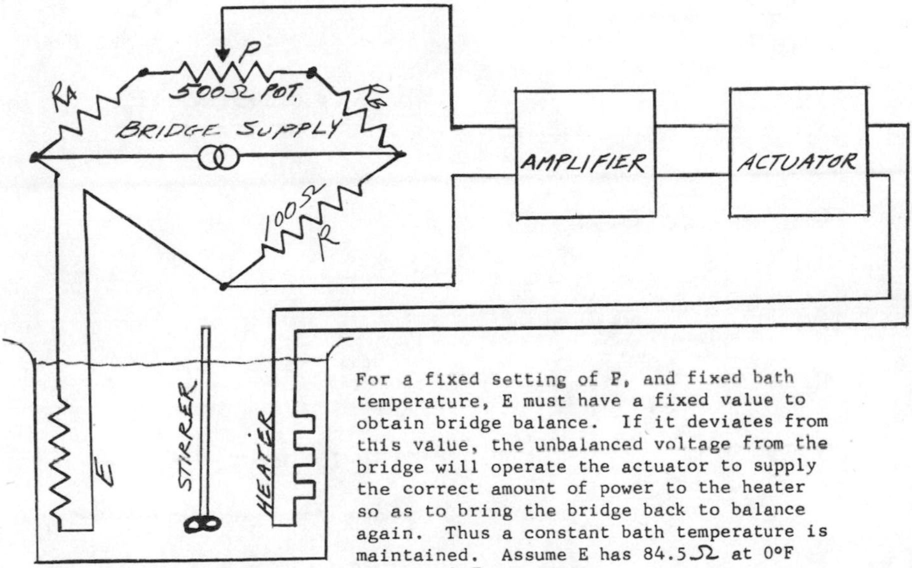

For a fixed setting of P, and fixed bath temperature, E must have a fixed value to obtain bridge balance. If it deviates from this value, the unbalanced voltage from the bridge will operate the actuator to supply the correct amount of power to the heater so as to bring the bridge back to balance again. Thus a constant bath temperature is maintained. Assume E has 84.5 Ω at 0°F and 214.6 Ω at 500°F.

REQUIRED:

Calculate the values of R_A and R_B so that by adjusting P, the bath temperature can be set at anywhere between 0° and 500°F.

This involves solving a typical bridge balance problem with potentiometer "P" set to either of its extremes:

For 0° temperature

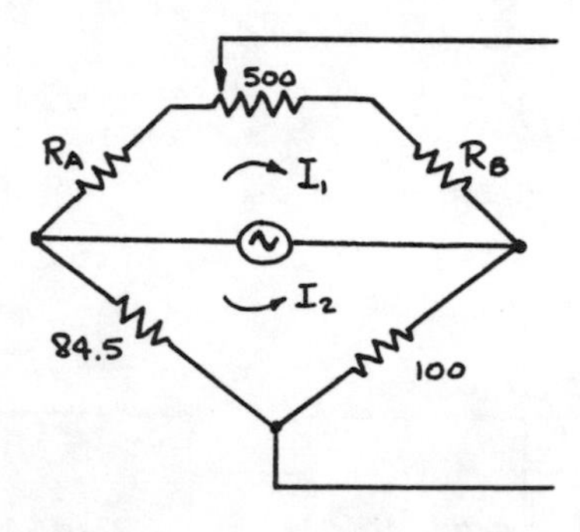

$$I_1 R_A = 84.5\, I_2$$

and

$$I_1(500 + R_B) = 100\, I_2$$

dividing one equation by the other

$$\underline{R_A = 0.845(500 + R_B)} \qquad (1)$$

And for 500° temperature

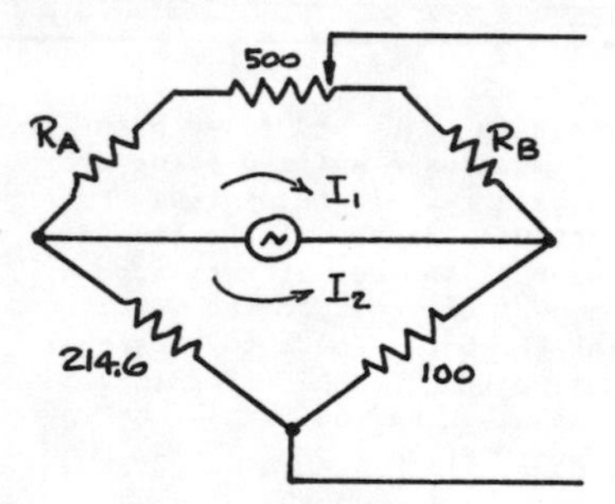

$$(500 + R_A) I_1 = 214.6\, I_2$$

and

$$R_B I_1 = 100\, I_2$$

dividing one equation by the other gives

$$\underline{500 + R_A = 2.146\, R_B} \qquad (2)$$

Then solving these two simultaneous equations (1 & 2) for R_A and R_B gives:

$$R_B = 710\ \Omega$$

$$R_A = 1022\ \Omega$$

CONTROL SYSTEMS 4.

A viscous-damped servomechanism with proportional control stiffness K and damping coefficient B, is found to exceed the allowing tracking error by 10 times.

The latter is corrected by altering the stiffness, and by adding derivative control to provide the original degree of underdamping.

REQUIRED:

Find the new proportional control stiffness and the derivative action time required.

A simple proportional control servomechanism might be represented as:

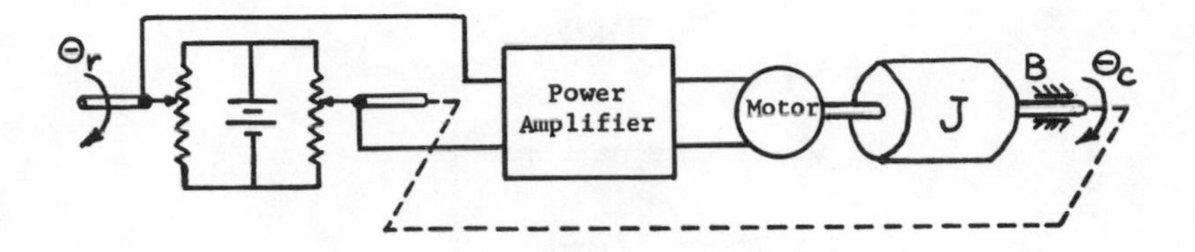

The equation describing this system could be written as:

$$K(\Theta_r - \Theta_c) = J\ddot{\Theta}_c + B\dot{\Theta}_c$$

(Equation assumes an oversimplified relationship where the motor torque is proportional to input voltage.)

And the Laplace transformed system then could be shown to be:

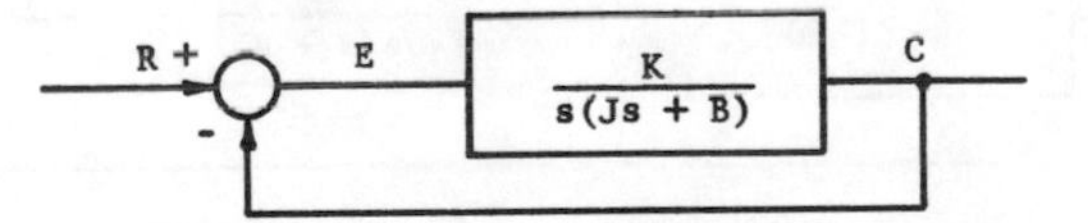

For ease of computing, let J = 1, then the system transfer function becomes:

$$G_{syst} = \frac{K}{S^2 + Bs + K} \equiv \frac{\omega_n^2}{S^2 + 2\zeta\omega_n S + \omega_n^2}$$

where $\omega_n = \sqrt{K}$

$$\zeta = \frac{B}{2\sqrt{K}}$$

The system tracking error (assuming a unit ramp input) then may be found using the final value theorem:

$$e\Big|_{t\to\infty} = \lim_{s\to 0} sE = \lim_{s\to 0} s\left[\frac{R}{1+G}\right] = \lim_{s\to 0} s\left[\frac{\frac{1}{s^2}}{1+\frac{K}{s(s+B)}}\right] \to \frac{B}{K}$$

And, the root locus will then be (assuming an underdamped system):

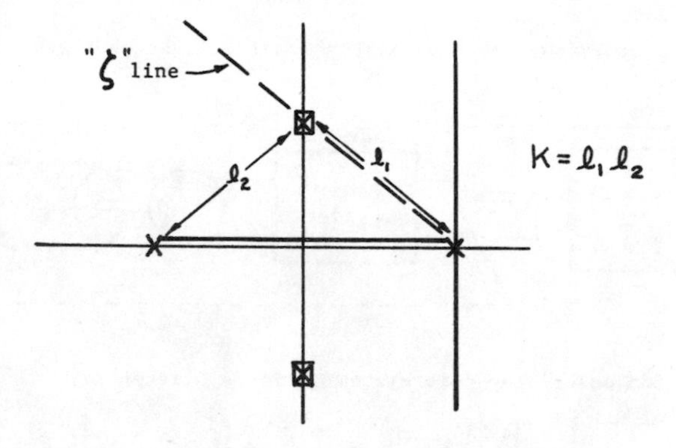

To reduce the system error to 0.1 of the original error, add an ideal derivative compensator:

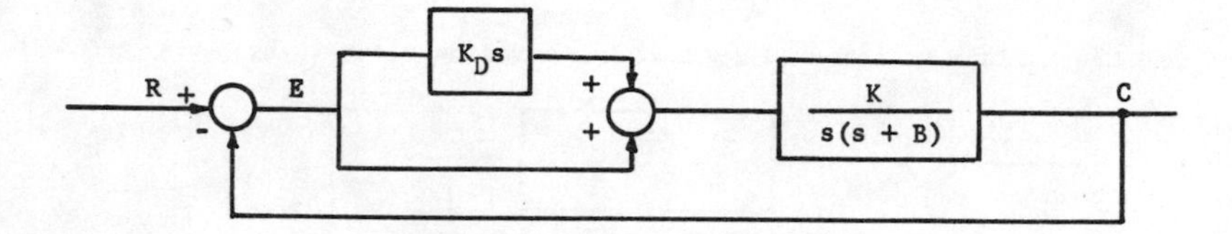

The new forward transfer function then becomes:

$$G_{fwd} = \frac{(1 + K_D s)K}{s(s + B)}$$

And the error again may be found using the final value theorem:

$$e\Big|_{t\to\infty} = \lim_{s\to 0} s\left[\frac{\frac{1}{s^2}}{1+\frac{(1+K_D s)K}{s(s+B)}}\right] \to \frac{B}{K}$$

To reduce the new error by a factor of 10 by increasing the stiffness, K, gives:

$$e_{new} = 0.1 e_{old} = \frac{B}{10K}$$

The new root locus will then have a zero located at $-\frac{1}{K_D}$ (yet to be determined).

Since the amount of damping is to remain the same (i.e., the "ζ" line to remain the same), the locus will be of the form:

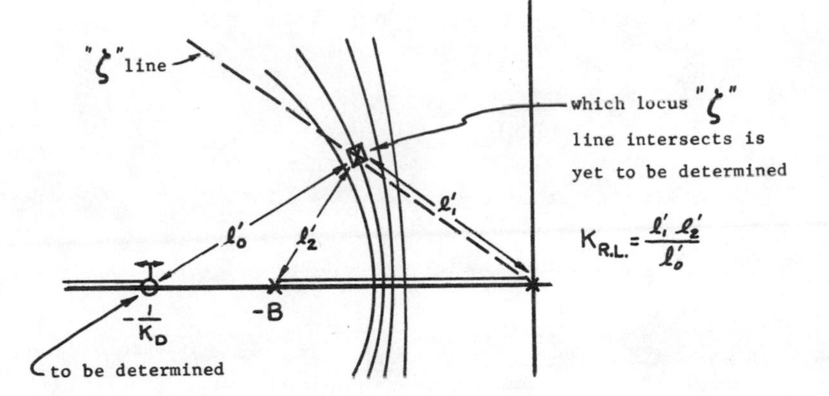

The requirement of error, being $K_{new} = 10 \times K_{original}$, and locating the new closed loop pole on the "ζ" line will give the relationship of:

$$K_{root\ locus} = \frac{\ell_1' \ell_2'}{\ell_0'} = K_D(10K) = K_D 10 \ell_1 \ell_2$$

Using this relationship, one may, by trial and error techniques, find the value of K_D.

However, a more straight forward technique not using root locus methods, will allow us to approximate K_D directly (this will be approximate as the zero in the closed loop response will alter the damping).

Consider:

$$G_{syst} = \frac{(10K)(1 + K_D s)}{s^2 + Bs + K_D(10K)s + (10K)} \equiv \frac{\omega_n^2(1+K_D s)}{s^2+2\zeta\omega_n s+\omega_n^2}$$

where $\omega_n = \sqrt{10K}$

$$2\zeta\omega_n = B + 10K\,K_D$$

$$\zeta = \frac{B+10K\,K_D}{2\sqrt{10K}}$$

But the new "ζ" and original "ζ" are required to be the same.

Giving:

$$B = \frac{B + 10K\,K_D}{\sqrt{10}}$$

As an example, let K = B = 2 and "ζ" = 0.707

Then

$$2 = \frac{2+(10)(2)K_D}{\sqrt{10}}$$

Giving K_D = 0.216

Or, the zero (for the root locus plot) is located approximately at $\frac{-1}{K_D}$, which gives -4.6 for the numbers previously used.

Therefore the new stiffness should be 10 times the original and the "time constant" of the derivative compensator should equal K_D.

CONTROL SYSTEMS 5.

A servo system for the positional control of a rotatable mass is stabilized by means of viscous damping. The amount of damping is less than that required for critical damping.

REQUIRED: Calculate the amount (percent) of the first overshoot, if the input member is suddenly moved to a new position, the undamped natural frequency is 5 Hz, and the viscous friction coefficient is a fifth of that required for critical damping.

Solution

Assume an ideal controller with torque directly proportional to error, E, (T = KE), and that system is of 2nd order:

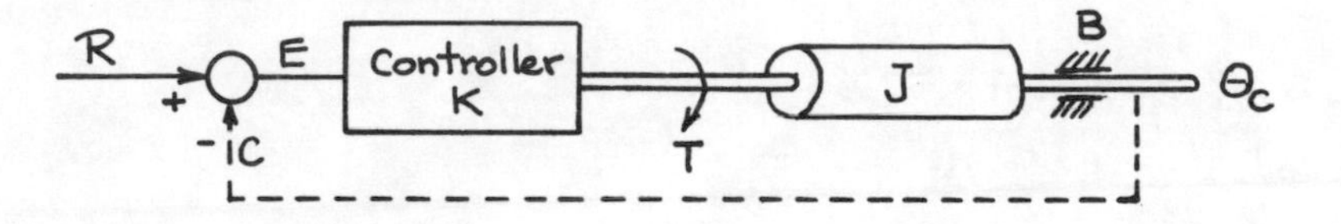

$$T = KE = K(R-C)$$

$$T = (JS^2 + BS)\Theta$$

$$G_{SYST} = \frac{K}{JS^2 + BS + K} = \frac{K/J}{S^2 + \frac{B}{J}S + K/J}$$

$$= \frac{\omega_n^2}{S^2 + 2\zeta\omega_n S + \omega_n^2}$$

ω_n = natural frequency = $2\pi f_n = 2\pi 5$ = constant

$2\zeta\omega_n = B/J$, $\therefore \zeta \propto B$ since both K and J are constants.

For critical damped system, $\zeta = 1$.

For $B = 1/5\, B_{critical}$, $\zeta = 0.2$

From "standardized" 2nd Order Curves:

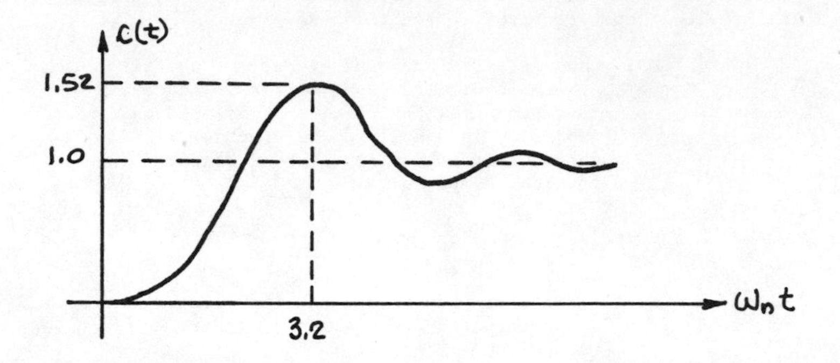

∴ % overshoot = 52%

and time-to-first-peak, t_p

$$t_p \approx \frac{3.2}{\omega_n} \approx 0.1 \text{ seconds}$$

8

ILLUMINATION

ILLUMINATION 1.

In the planning of the lighting for a standard classroom for a junior high school, three schemes of lighting equipment are under consideration as follows:

a. Eight - 500 watt pendent mounted incandescent lighting fixtures with semi-indirect plastic diffusers.
b. Three rows of two-lamp ceiling mounted fluorescent strip fixtures giving a total of 21 four-foot fixtures.
c. Two rows of three-lamp suspended fluorescent fixtures with metal louvers giving a total of 14 four-foot fixtures.

(1) Wt. 6 What are three or more advantages and three or more disadvantages of each scheme?

(2) Wt. 4 Which scheme would you recommend and why? Or, you may prefer to propose a fourth scheme as your recommendation. If so, state why.

Incandescent lighting plan (a)

8 - 500 watt incandescent @ 10,000 lumens= 80,000 lumens

8 - 500 watt incandescent = 4000 watts

Fluorescent lighting plan (b)

4 foot - 40 watt white fluorescent lamps
21 fixtures x 2 = 42 lamps

42 - 40 watt lamps @ 2000 lumens = 84,000 lumens

42 - 40 watt lamps =	1680 watts
21 fixture ballasts @ 20 watts =	420 "
	2100 watts

Fluorescent lighting plan (c)

14 fixtures x 3 = 42 lamps

42 - 40 watt lamps @ 2000 lumens = 84,000 lumens

42 - 40 watt lamps =	1680 watts
14 fixture ballasts @ 20 watts =	280 "
	1960 watts

(1)

Plan (a) Advantages:

1. Initial cost will be less
2. Less fixtures and lamps to maintain
3. No stroboscopic effect

Plan (a) Disadvantages:

1. 4000 watts will be dissipated in the room
2. Illumination will not be uniform across the room. Intensity will be high under the lights and low between lights.
3. Efficiency will go down as ceiling and walls get dirty lowering the coefficient of utilization. Reflection factor plays an important role in this installation

Plan (b) Advantages:

1. This plan presents the most uniform lighting of the three plans.
2. Heat dissipation is 2100 watts
3. Fixtures mounted on the ceiling means that the light will all be diffused downward so less light will be absorbed in the ceiling.

Plan (b) Disadvantages:

1. Initial cost will be more
2. More lamps and fixtures to maintain
3. Stroboscopic effect bothers some people

Plan (c) Advantages:

1. Three lamp fixtures reduce stroboscopic effect
2. Three lamp fixtures may have white and daylight tubes to give better visibility curves. This is also true of the two tube fixtures.
3. Not as costly as plan (b) for initial installation and plan (c) takes the least amount of energy

Plan (c) Disadvantages:

1. Distribution will not be uniform across the room.
2. This design will be inefficient in using the advantages available in room index.
3. This plan would have a lower coefficient of utilization than plan (b).

(2) Scheme recommended.

Plan (b) is the most desirable unless the ceilings are too high, in which case the same scheme should be suspended from the ceiling at the proper height. The advantages were given as follows:

1. Less energy dissipated in the room
2. Lighting will be more uniform than that produced by the other two plans.
3. Saving in power consumption will be substantial over a period of time
4. Instant start fluorescent tubes are long life and the fixtures are nearly trouble free.
5. Fluorescent lighting in a school plant reduces power cost by a considerable amount.

The ideal light is an infinite point source or an illuminated ceiling. This is too costly for standard classrooms. The life of fluorescent lamps is about 2,000 hours, twice the life of incandescent lamps.

ILLUMINATION 2.

(a) Fluorescent lamps come in a wide variety of lengths, current ratings, and types. Of the following, select the four which are clearly superior to the rest in providing the most light for the least money when used in two-lamp industrial fixtures:

F8T5	F20T12	F72T12	F96T12/HO
F15T8	F30T12	F96T12	F48PG17
F30T8	F40	F48T12/HO	F72PG17
F96T8	F48T12	F72T12/HO	F96PG17

(b) What levels of illumination are currently recommended by the Illuminating Engineering Society for these tasks:

1. General classroom work
2. General office work
3. Rough layout drafting
4. Medium bench and machine work
5. Fine assembly
6. Intermittent filing
7. Corridors
8. Auto body finishing

(c) When a lighting level of 500 foot-candles is to be obtained by a combination of general and supplementary lighting, what should be the minimum level of general lighting provided?

(d) Give the three major reasons for putting apertures in the top of the reflectors of fluorescent industrial luminaires.

(e) 1. Name the three types of starting circuit commonly used for fluorescent lamps 48" or more in length.
2. Which of the three is most commonly used?
3. Which is practically obsolete?

(f) Give three commonly used permanent, nonadjustable methods of controlling the brightness of classroom window areas in school construction.

(g) Fluorescent industrial fixtures are installed 10 feet above the work plane. For uniform illumination, approximately what maximum spacing of fixture rows should be used?

(h) Approximately what maintained level of illumination can be expected from two-lamp 800-ma industrial units installed in continuous rows on 10' centers?

(i) What approximate efficiency can be expected from well-designed fluorescent luminaires of the following types:

1. Recessed 2 x 4 with prismatic panels
2. Direct-indirect commercial units with cross baffles
3. Industrial fixtures with apertures for about 15% uplight
4. Recessed 1 x 4 with metal 1/2" cube egg-crate louvers
5. Luminous indirect
6. Ceiling-mount "Wrap-around" with top plates

(j) A photometric test report normally contains the following data. For each item, discuss the use which may be made of it in design:

1. Candlepower distribution curve
2. Maximum brightness readings
3. Average brightness values (or curves)
4. Coefficients of utilization

(a) Of the standard industrial fixtures:

1. F96PG17 3. F96T12/HO
2. F96T12 4. F40

(b) 1. 70 f.c. 5. 500 f.c.
2. 100 f.c. 6. 70 f.c.
3. 150 f.c. 7. 20 f.c.
4. 100 f.c. 8. 200 f.c.

(c) 100 f.c.

(d) 1. Act like a chimney to produce a draft.
2. Provide some lighting on the ceiling (reduces dungeon effect).
3. Reduce the luminescence ratios between fixture and background.

(e) 1. Rapid start, lead-lag, preheat.
2. Rapid start.
3. Preheat.

(f) 1. Overhangs (roof)
2. Louvers
3. Landscaping (trees)

(g) Depending on luminaire distribution we can develop the following set of ratios from spacing (ft)/room height (ft).

-0.5, highly concentrated; 0.5 - 0.7, concentrated; 0.7 - 1.0, medium spread; 1.0 - 1.5, spread; over 1.5, widespread.

(h) The level of illumination can be calculated as follows:

$$\frac{8900 \text{ lumens} \times 2 \text{ lamps}}{10 \text{ ft} \times 8 \text{ ft}} \times 0.6 \times 0.65 = 87 \text{ f.c.}$$,where C.U.=0.6 and M.F.=0.65

(i) 1. 75% 4. 40 - 50%
2. 70 - 79% 5. 65 - 70%
3. 84% 6. 65 - 70%

(j) 1. Indicates the distribution of the light and therefore it is used to tell the spacing of the fixtures.
2. Used in the selection of fixtures where brightness is a problem or requirements have been set at specific fixture viewing angles.
3. It is used for the selection of fixtures to meet recommended brightness criteria such as set up in the scissor curve graph for school and office lighting.
4. In effect indicates the efficiency of a fixture in a specific room (as a function of: certain room size, certain room reflectance and certain luminaire).

Reference: IES Handbook.

ILLUMINATION 3.

PART A

Wt.

4 Three fluorescent lighting systems are considered for a 40'x60' general office with a 10' ceiling. Reflectances are: Ceiling 80%, walls 50%, floor cavity 20%. The systems are:

(1) Recessed 4-lamp 2x4 units with flat prismatic lenses.

(2) A diffusing luminous ceiling utilizing strips in a sealed, high-reflectance plenum.

(3) Direct-indirect luminaires suspended 24" from ceiling, with low-brightness side panels.

All luminaires are high quality, well designed units. All use the same lamp type.

REQUIRED:

(a) Which system would be selected for each of the following qualities?

1. Maximum illumination per watt of power used.
2. Best lighting for drafting boards.
3. Lowest maintenance required.
4. Coolest operation of lamps and ballasts.
5. Lowest initial luminaire cost for equal illumination.

(b) Which system would have the lowest visual comfort rating at high foot-candle levels (200 plus)?

(c) Give one practical way of improving the visual comfort of system (1) without changing the illumination level greatly.

(d) Give two practical ways of improving the visual comfort of system (2) without changing the illumination level greatly.

(e) The architect suggests using top plates on the luminaires of system (3) and ceiling mounting them to improve the appearance of the system. Discuss the effect of such a change on:

1. Visual comfort.
2. Reflected glare.
3. Illumination level.

PART B

Wt.

3 A room 14'x16', with a 10' ceiling, has reflectances of 80% ceiling, 50% wall, and 20% floor cavity. Sketch in detail a lighting system which will give as close as possible to uniform illumination over the entire room. Give enough detail on the lighting unit so that its characteristics are clear, including a sketch of the distribution curve.

PART C

Wt.

3 The following luminaire description appears in a specification. Several errors have been made in it - errors being defined as impossibilities, equipment not available, or mistakes in description. Write down each phrase in error, and explain why it is wrong:

> Luminaire shall be recessed, nominal 2'x4' in size, drop-in type, for mounting in inverted T ceiling. Luminaire length shall not exceed 48", for continuous row mounting. Luminaire shall contain four 40-watt rapid-start lamps, rated at 3100 lumens each, and shall be not less than 60% in efficiency. Average brightness shall not exceed 500 foot-lamberts at any angle from zero degrees (nadir) to 90°.
>
> Ballast shall be two-lamp, A-rated, 800-ma lead-lag type, with U/L and CBM labels. Ballast shall be protected by automatic resetting thermal protectors in series with primary, GE "Bonus Line" or equal. Lamp sockets shall be bi-pin, telescopic "push-pull" type.
>
> Luminaire shall be constructed of 28-gauge sheet steel, so formed that a continuous wireway is obtained through a row of luminaires.

PART A

(a) 1. No. 3: from the table for coefficients of utilization, No. 3 has a higher coefficient (for same height) than No. 1. In addition No. 3 is suspended 24" from the ceiling, therefore 2 feet closer to the work plane.

2. No. 3: source is closer to the work plane. Requirements for drafting are: Detailed drafting - 200 f.c.; rough layout drafting - 150 f.c.

3. No. 3: Fixture being suspended from the ceiling, it can be maintained much easier; a recessed fixture is built into the ceiling and therefore receives less regular maintenance.

4. No. 3: a suspended fixture receives regular ventilation and therefore it will have the coolest operation.

5. No. 3: having maximum illumination per watt of power used (see above part 1) this system will need fewer fixtures since for equal illumination less power is needed.

NOTE: Closeness to the work plane is not a clear cut advantage. The loss of light sent up to the ceiling and bounced back could outweigh the advantage of lowering the fixture.

(b) No. 3 system: since the other systems are not suspended from the ceiling, their light is more diffused.

(c) One way: select lenses with greater cutoff in the normal viewing angles; another way: substitute opaque plastic diffusers for glass lenses.

(d) 1. Select lenses which have greater shielding;
2. Place shielding baffles on the T-bars.

(e) 1. Visual comfort is decreased as now only the bottom side of the fixtures serves as light emitting and thus system will be brighter in contrast with the ceiling.
2. The contrast of dark ceiling and brighter fixtuer will accentuate glare.
3. Illumination level will remain about the same or could improve slightly depending on the reflectance of the ceiling; but the uniformity of the light in the room would be reduced.

PART B

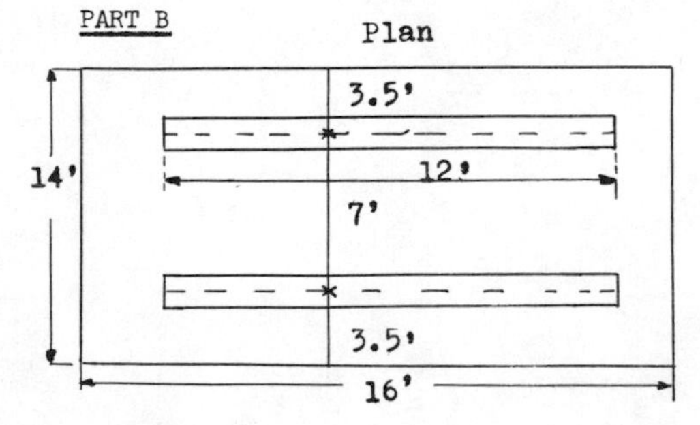

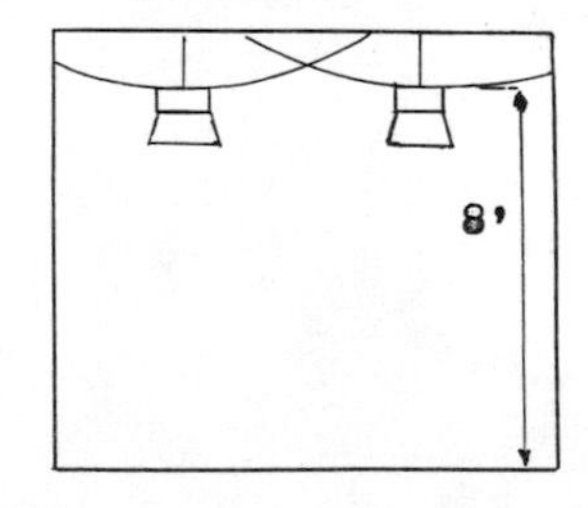

Foregoing system will give uniform illumination and will be relatively inexpensive. Here high output (HO) or very high output (VHO) lamps can be considered.

PART C

1. "Luminaire shall be recessed, nominal 2'x4' in size, drop-in type, for mounting in inverted T ceiling. Luminaire length shall not exceed 48", for continuous row mounting."
   a. Conflict with the lay-in feature of the fixture: can't fit in 4' fixture in continuous row mounting due to narrow working clearance at ends close to walls.
   b. One can't "drop in" a luminaire to an inverted ceiling T bar.
2. "Luminaire shall contain four 40-watt rapid start lamps, rated at 3100 lumens each, and shall be not less than 60% in efficiency"... Ballast shall be two-lamp, A-rated, 800 -ma.... type"
   a. 40-watt RS lamp is only rated for 430 ma and does not operate on an 800 ma ballast.
   b. The quoted efficiency of 60% is way too high.
3. "Average brightness shall not exceed 500 foot-lamberts at any angle from zero degrees (nadir) to 90°
   a. To achieve the brightness from zero degrees to 90° is virtually impossible, especially around 0°.
4. "Ballast shall be ....lead-lag type"...Luminaire shall contain four 40-watt rapid-start lamps"
   a. Lead-lag type ballast and rapid-start lamps are two completely different systems and do not work together.
5. "Ballast shall be protected by automatic resetting thermal protectors in series with primary GE "Bonus line" or equal."
   a. Automatic resetting feature is exactly the opposite of the "Bonus line" system, the latter being a non-automatic resetting fixture; therefore the quoted luminaire cannot have both features at the same time.
6. "Lamp sockets shall be bi-pin, telescopic "push-pull" type."
   a. It is impossible for a "push-pull" type socket to meet the 48" overall length requirement with 430 ma rapid-start lamp.
7. "Luminaire shall be constructed of 28-gauge sheet steel"
   a. This gauge size is too thin for rigidity. Usually #20 ga is used for housing and #18 for brackets to hold lampholders, etc. Even the 20 ga is formed with ribs for rigidity.
8. "Luminaire....so formed that a continuous wireway is obtained through a row of luminaire"
   a. The continuous wireway feature conflicts with the lay-in feature of the system.

Reference: IES Lighting Handbook

ILLUMINATION 4.

REQUIRED:

Wt.

1 (a) Name six basic types of lamp light sources available on a commercial basis.

1 (b) Of these types of lamps available, which would you most commonly use to light a

1. residence
2. school
3. gymnasium
4. football field
5. highway
6. department store
7. radio frequency shielded room

1 (c) If a lighting fixture salesman assures you that his fixtures are absolutely the best available for the price, what items of information concerning his fixtures would you ask him to give you to verify his statements?

3 (d)
1. What is the name of the unit of light source intensity?

2. For a given point light source 10 feet from a flat illuminated surface, what is the theoretical ratio of illumination received to illumination produced?

3. If a classroom is lighted to a general level of 75 foot-candles, and the fixtures are mounted 8 feet above the floor, what light level would you expect if, for structural reasons, all the fixtures had to be raised 2 feet?

2 (e)
1. The United States armed forces years ago used high intensity lights as aircraft search lights. What was the light source in the search lights?

   Which factors caused the Armed Forces to discontinue the use of these lights? (Name four)

2. Since light is electro-magnetic energy, and the old search lights had a high light output, is it likely that they could be used as detection elements of radar systems? Why?

2 (f)
1. Highways are lighted at some places for driver and pedestrian safety. 400-watt lamps are usually used. Would you, personally, favor using 1000-watt lamps and fixtures on the same poles for increased illumination?

2. Could increasing the light source brightness actually increase the hazard at such crossings? If so, how would you correct the situation?

(a)
1. Incandescent (including quartz iodine type lamp light source); 2. Fluorescent; 3. Mercury vapor; 4. Short arc; 5. Glow lamp; 6. Photo flash.
(b)
1. Incandescent; 2. Fluorescent; 3. Mercury vapor; 4. Mercury vapor and/or quartz iodine; 5. Mercury vapor; 6. Fluorescent; 7. Fluorescent with shielded luminaire (namely sensitive electronic equipment can receive stray signals emitted by the fluorescent fixture; shielded luminaires minimize such emissions).
(c)
1. Tables denoting "coefficients of utilization"; 2. Other photometric data such as brightness tables or curves; 3. Candlepower distribution curves; 4. Glare factor tables; 5. Material of the fixture - type, thickness, etc.; 6. Examine a fixture personally.
(d)
1. Candlepower

2. 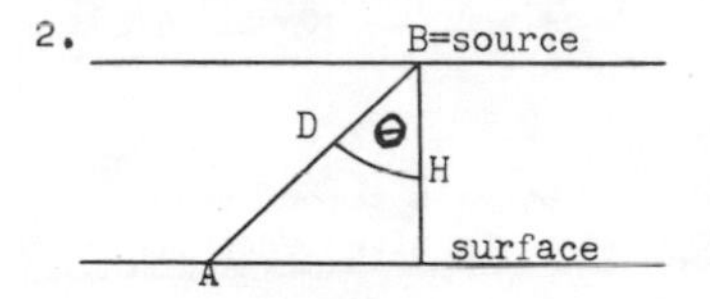

$$FC = \frac{CP}{D^2} = \frac{CP}{100}$$ where:

FC = illumination of the surface or footcandles at point A
CP = Luminous intensity in candlepower
D = distance, here given as 10 feet

3. The floor area remains the same and the increase of illuminated wall area (due to the raising of the fixtures by 2 feet) will reduce the amount of light reflected to the floor by the wall's reflectance; but will not significantly affect the footcandle level at the floor. This is a slight reduction, nowhere near the result if the square-law were directly applied. The square law namely states that footcandles measured are directly proportional with the square of the distance between the source and the point of measurement, within a certain range.
(e)
1. Carbon arc light source was used in the search lights.
Factors which caused the discontinuance of these lights:
I. Difficult to maintain; II. Short electrode life; III. Complex mechanism; IV. Cannot be left unattended for any length of time.
2. Although both (light and radar) can be considered as waves, the characteristics of the much shorter radar wavelength enables it to penetrate barriers, such as clouds, and then be reflected to a receiver. The much longer wavelengths of light have this property to a very lesser degree.
(f)
1. Not necessarily, because using 1000-watt lamps in lieu of 400-watt lamps may cause "hot" spots which would increase light to dark ratios, thus making items outside the light more difficult to see.
2. Yes, increased brightness would increase the hazard at road crossings. The situation can be corrected by increasing the mounting height of the light source. If 100-watt light source were used at increased mounting heights, this would decrease the brightness ratio at a greater distance from the luminaire located at the crossing.

Reference: IES Lighting Handbook

ILLUMINATION 5.

The following problem concerns lighting design, and answers should be based on the latest editions of applicable references:

(a) Give the equation for calculating the cavity ratios, and name the ratios required for a room illuminated with pendent fixtures.

(b) What two values of reflections are adjusted by use of the foregoing ratios?

(c) A maintenance factor is normally used in interior lighting calculations. Give three main reasons why. (Neglect wiring)

(d) A classroom for a high school 24 ft x 32 ft x 10 ft high is to be illuminated, using recessed fluorescent fixtures with low brightness acrylic prismatic lenses and two 40-watt RS (3100 Lu) lamps per fixture. Compute the number of fixtures required assuming a CU of 0.6.

What would be an appropriate M.F.? What is a recommended footcandle level? How many fixtures are required theoretically?

(e) Sketch the actual layout showing the fixture spacing and distance to walls, for the room in Part (d). Assume the front of the room is at the 24-ft end.

What is the computed maintained footcandle level of the layout shown?

Answers

(a) In the Zonal - Cavity Method the effects of room proportions, luminaire suspension length, and work plane height upon the coefficient of utilization are respectively accounted for by the Room Cavity Ratio, Ceiling Cavity Ratio, and Floor Cavity Ratio.

These ratios are determined by dividing the room into three cavities as shown below and substituting dimensions (in feet) in the following formula.

$$\text{Cavity Ratio} = \frac{5 \cdot h\ (\text{Room Length} + \text{Room Width})}{\text{Room Length} \times \text{Room Width}}$$

where, $h = h_{RC}$ for the Room Cavity Ratio, RCR

$= h_{CC}$ for the Ceiling Cavity Ratio, CCR

$= h_{FC}$ for the Floor Cavity Ratio FCR

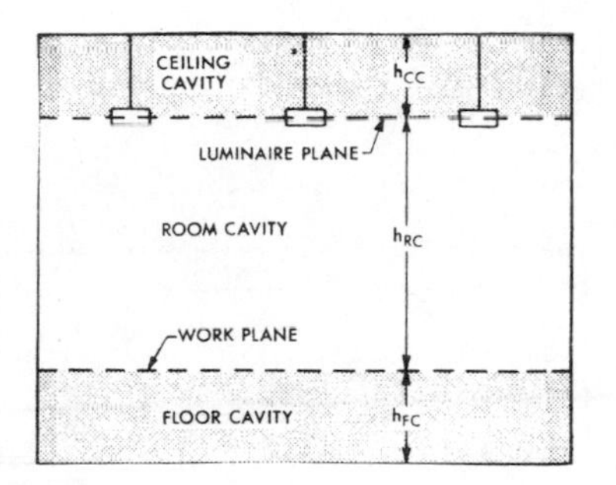

Note that $CCR = RCR \dfrac{h_{CC}}{h_{RC}}$

and $FCR = RCR \dfrac{h_{FC}}{h_{RC}}$

(b) By the use of the foregoing ratios the combination of wall and ceiling or wall and floor reflections are adjusted. We can convert the combination of wall and ceiling or wall and floor reflections into a single Effective Ceiling Cavity Reflectance and a Single Effective Floor Cavity Reflectance.

(c) A maintenance factor is normally used in interior lighting calculations because of three main reasons.

1. accumulation of dirt on room surfaces.
2. lamp lumen depreciation.
3. luminaire dirt.

Additional reasons could be

4. temperature
5. aging of luminaire finish and ceiling material.

(d) The number of fixtures required can be obtained as follows:

$$\frac{LF \times N \times CU \times MF}{W \times L} \times Z = FC$$

where:

LF is the luminous flux in lumens
N is the number of lamps per fixture
CU is the coefficient of utilization (fixture and room effects on illumination)
MF is the maintenance factor (see (c) above for definition)
W,L are the width and length of room
Z is the number of fixtures
FC is the illumination or foot candle level

For a classroom, the IES recommends about 75 FC.

Using MF = 0.75 and N = 2 we obtain:

$$\frac{3100 \times 2 \times 0.6 \times 0.75}{24 \times 32} \times Z = 75$$

$$3.63 \times Z = 75$$
$$Z = 21$$

(e)

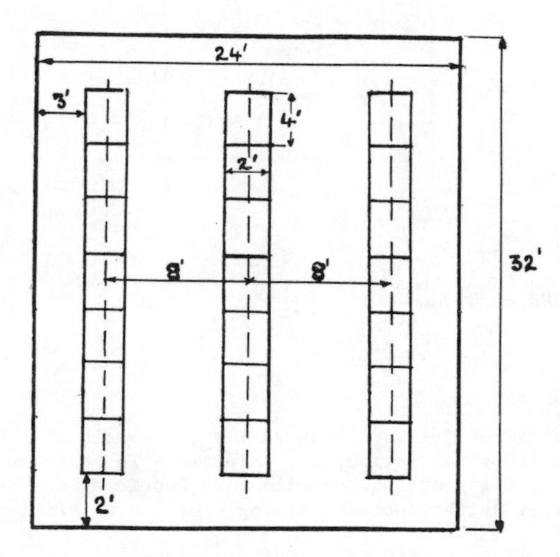

9
ECONOMIC ANALYSIS

Engineering economic analysis (often called engineering economy) is a group of techniques for the systematic analysis of alternative courses of action. This chapter will begin with a detailed review of economic analysis fundamentals, including

- Notation
- Cash Flow
- Time Value of Money
- Equivalence
- Compound Interest

An understanding of these fundamentals is prerequisite to the discussion of economic analysis techniques needed for the electrical engineering examination. The techniques are grouped as follows:

- Present Worth
- Annual Cost
- Rate of Return
- Depreciation & Taxes

Examination problems and solutions follow this review of fundamentals and discussion of analysis techniques.

NOTATION

In the past there was considerable variation in economic analysis symbols and notation. Recent attempts at standardization have helped to reduce this variation.

Symbols

	Obsolete Symbol	Standard Symbol
Present Sum	P	P
Future Sum	S	F
End-of-period payments or receipts in a uniform series continuing for a specified number of periods	R	A
Number of interest periods*	n	n
Interest rate per interest period	i	i

*The standard symbol is actually N, but many authors continue to use the lower case n. Lower case n will be used here.

Compound Interest Factors

	Given	To Find	Obsolete Mnemonic	Standard Mnemonic	Standard Functional
Single Payment					
Compound Amount Factor	P	F	(caf'-i%-n)	(CA-i%-n)	(F/P,i%,n)
Present Worth Factor	F	P	(pwf'-i%-n)	(PW-i%-n)	(P/F,i%,n)
Uniform Series					
Sinking Fund Factor	F	A	(sff-i%-n)	(SF-i%-n)	(A/F,i%,n)
Capital Recovery	P	A	(crf-i%-n)	(CR-i%-n)	(A/P,i%,n)
Compound Amount Factor	A	F	(caf-i%-n)	(SCA-i%-n)	(F/A,i%,n)
Present Worth Factor	A	P	(pwf-i%-n)	(SPW-i%-n)	(P/A,i%,n)

In this book we will use the standard functional notation.

CASH FLOW

In examining alternative ways of solving a problem we recognize the need to resolve the various consequences (both favorable and unfavorable) of each alternative into some common unit. One convenient unit - and the one typically used in economic analysis - is money. Thus an initial step in resolving economic analysis problems is to convert the various consequences of an alternative into a table of year-by-year cash flows.

For example, a simple problem might be to portray the consequences of purchasing a new car as follows:

	Year	Cash Flow	
Beginning of first year	0	-4500	Car purchased "now" for $4500 cash. The minus sign indicates a disbursement.
End of year	1	-350	Maintenance costs are $350 per year.
End of year	2	-350	
End of year	3	-350	
End of year	4	-350 +2000	The car is sold at the end of the 4th year for $2000. The plus sign represents a receipt of money.

This same cash flow may be represented graphically as is shown on the top of the next page.

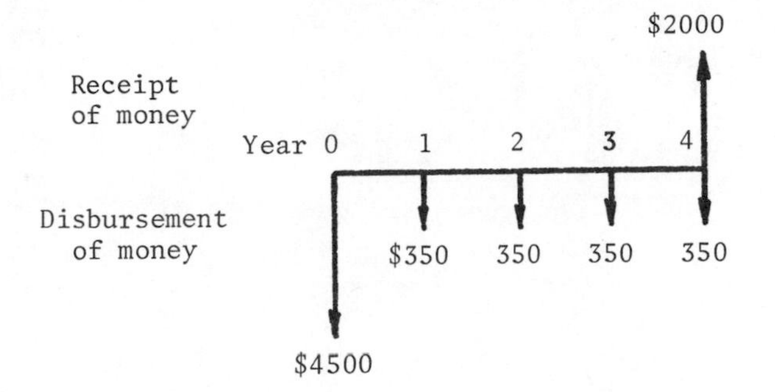

The upward arrow represents a receipt of money, and the downward arrows represent disbursements. The X-axis represents the passage of time.

TIME VALUE OF MONEY

When the money consequences of an alternative occur in a short period of time - say less than one year - we might simply algebraically add up the various sums of money and obtain the net result. But we cannot treat money this same way over longer periods of time. This is because money today is not the same as money at some future time.

Consider this question. Which would you prefer, $100 today or the assurance of receiving $100 a year from now? Clearly you would prefer the $100 today. If you had the money today, rather than a year from now, you could use it for the year. And if you had no use for it, you could lend it to someone who would pay interest for the privilege of using your money for the year. Thus $100 today would be equivalent, at 5% interest, to $105 a year hence.

EQUIVALENCE

In the preceding section we saw that money at different points in time (e.g., $100 today or $100 one year hence) may be equal in the sense that they both are $100, but $100 a year hence is not an acceptable substitute for $100 today. When we have acceptable substitutes, we say they are <u>equivalent</u> to each other. Thus at 5% interest $105 a year hence is equivalent to $100 today.

Equivalence is an essential factor in engineering economic analysis. Suppose we wish to select the better of two alternatives. First, we must compute their cash flows. An example would be:

	Alternative	
Year	A	B
0	-2000	-2800
1	+800	+1100
2	+800	+1100
3	+800	+1100

The larger investment in Alternative B results in larger subsequent benefits, but we have no direct way of knowing if Alternative B is better than Alternative A. Therefore, we do not know which alternative should be selected. To make a decision, we must resolve the alternatives into equivalent sums so they may be compared accurately and a decision made.

COMPOUND INTEREST

To facilitate equivalence computations a series of compound interest formulas will be derived and used.

Single Payment Formulas

Suppose a present sum of money P is invested for one year at interest rate i. At the end of the year we should receive back our initial investment P together with interest equal to Pi or a total amount P+Pi. Factoring P, the sum at the end of one year is P(1+i). If we agree to let our investment remain for subsequent years, the progression is as follows:

	Amount at Beginning of Period	+	Interest for the Period	=	Amount at End of the Period
1st year	P	+	Pi	=	$P(1+i)$
2nd year	$P(1+i)$	+	$Pi(1+i)$	=	$P(1+i)^2$
3rd year	$P(1+i)^2$	+	$Pi(1+i)^2$	=	$P(1+i)^3$
nth year	$P(1+i)^{n-1}$	+	$Pi(1+i)^{n-1}$	=	$P(1+i)^n$

The present sum P increases in n periods to $P(1+i)^n$. This gives us a relationship between a present sum P and its equivalent future sum F.

$$\text{Future Sum} = (\text{Present Sum})(1+i)^n$$

$$F = P(1+i)^n$$

This is the single payment compound amount factor. In standard functional notation it is written:

$$F = P(F/P,i\%,n).$$

The relationship we have derived may be rewritten as

$$\text{Present Sum} = (\text{Future Sum})(1+i)^{-n}$$

$$P = F(1+i)^{-n}$$

This is the single payment present worth factor. In standard functional notation it is written:

$$P = F(P/F,i\%,n).$$

Uniform Series Formulas

Consider the following situation.

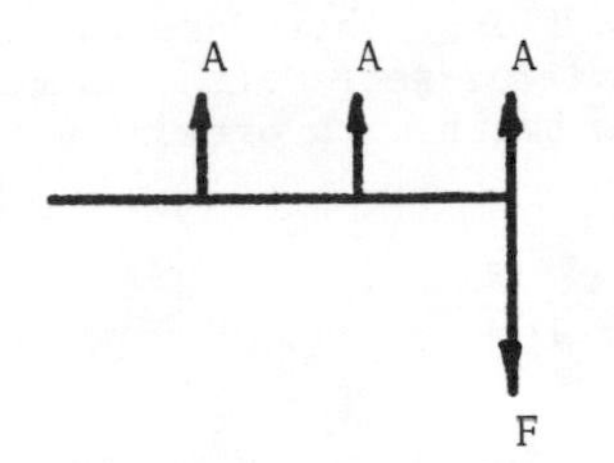

A = end-of-period receipt or disbursement in a uniform series continuing for n periods.
F = a future sum of money.

Using the single payment compound amount factor we can write an equation for F in terms of A.

$$F = A + A(1+i) + A(1+i)^2 \qquad (1)$$

In our situation, with n=3, Equation (1) may be written in a more general form:

$$F = A + A(1+i) + A(1+i)^{n-1} \qquad (2)$$

Multiply (2) by (1+i)

$$(1+i)F = A(1+i) + A(1+i)^{n-1} + A(1+i)^n \qquad (3)$$

Write Eqn(2)

$$F = A + A(1+i) + A(1+i)^{n-1} \qquad (2)$$

(3) - (2)

$$iF = -A + A(1+i)^n$$

$$F = A\left[\frac{(1+i)^n - 1}{i}\right]$$ Uniform Series Compound Amount Factor

Solving this equation for A

$$A = F\left[\frac{i}{(1+i)^n - 1}\right]$$ Uniform Series Sinking Fund Factor

Since $F = P(1+i)^n$ we can substitute this expression for F in the equation and obtain

$$A = P\left[\frac{i(1+i)^n}{(1+i)^n - 1}\right]$$ Uniform Series Capital Recovery Factor

Solving the equation for P

$$P = A\left[\frac{(1+i)^n - 1}{i(1+i)^n}\right]$$ Uniform Series Present Worth Factor

In standard functional notation the uniform series factors are:

Compound Amount	(F/A,i%,n)
Sinking Fund	(A/F,i%,n)
Capital Recovery	(A/P,i%,n)
Present Worth	(P/A,i%,n)

PRESENT WORTH

With the compound interest factors one may alter a cash flow into an equivalent sum or an equivalent cash flow. There are three major methods of comparing alternatives - present worth, annual cost, and rate of return. In this section we begin with present worth.

Criteria

Economic analysis problems inevitably fall into one of three categories:

1.	Fixed Input	The amount of money or other input resources are fixed. Example: A E.E. Project Engineer has a budget of $450,000 to overhaul a plant power system.
2.	Fixed Output	There is a fixed task, or other output, to be accomplished. Example: A mechanical contractor has been awarded a fixed price contract to air condition a building.
2.	Neither Input nor Output fixed	This is the general situation where neither the amount of money or other inputs, nor the amount of benefits or other outputs are fixed. Example: An E.E. consulting firm has more work available than it can handle. It is considering paying the staff for working evenings to increase the amount of design work it can perform.

In each of these categories we can determine what the criterion should be for economic efficiency. For present worth analysis the proper criteria are:

Category	Present Worth Criterion
Fixed Input	Maximize the present worth of benefits or other outputs.
Fixed Output	Minimize the present worth of costs or other inputs.
Neither Input nor Output fixed	Maximize [present worth of benefits minus present worth of costs] or stated another way: Maximize Net Present Worth.

Application of Present Worth Analysis

Present worth analysis is most frequently used to determine the present value of future money receipts and disbursements. We might want to know, for example, the present worth of an income producing property, like an oil well. This should provide a good estimate of the price at which the property could be bought or sold.

An important restriction in the use of present worth calculations is that there must be a common analysis period when comparing alternatives. It would be incorrect, for example, to compare the present worth (PW) of cost of Pump A, expected to last 6 years, with the PW of cost of Pump B, expected to last 12 years.

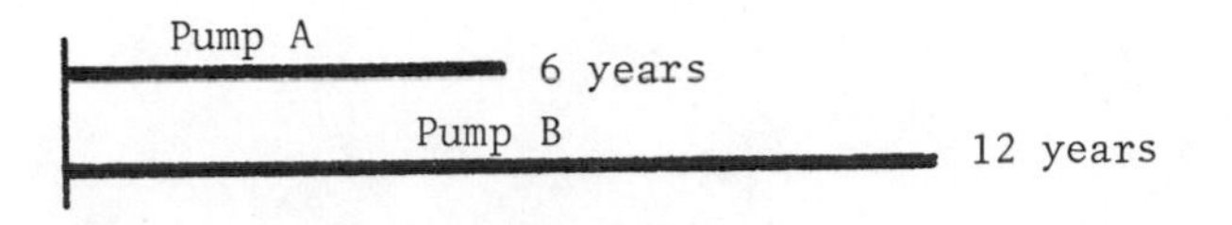

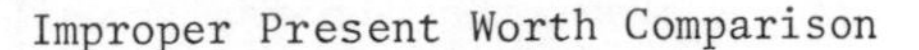
Improper Present Worth Comparison

In situations like this the solution is either to use some other analysis technique* or to restructure the problem so there is a common analysis period. In the example above, a customary assumption would be that there is a need for the pump for 12 years and that Pump A will be replaced by an identical Pump A at the end of 6 years. This gives a 12 year common analysis period.

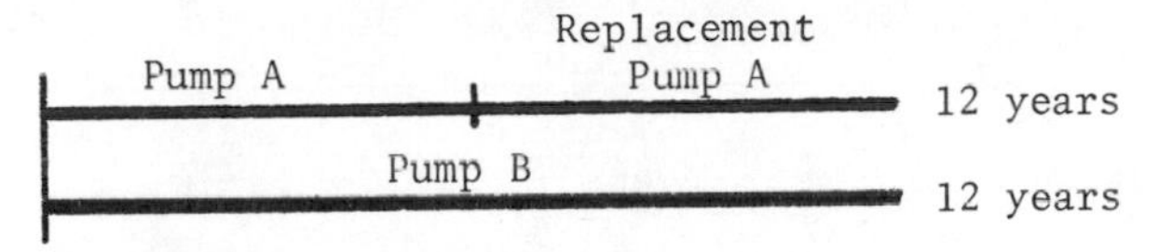

Correct Present Worth Comparison

This approach is easy to use when the different lives of the alternatives have a practical least common multiple life. When this is not true (e.g., Life X equals 7 years and Life Y equals 11 years), some assumptions must be made to select a suitable common analysis period or the present worth method should not be used.

Capitalized Cost

In the special situation where the common analysis period is infinite ($n = \infty$), an analysis of the present worth of cost is called capitalized cost. There are relatively few situations where the analysis period is infinity. In some public projects an infinite analysis period is used. Other examples would be permanent endowments and cemetery perpetual care.

---

*Generally the annual cost method is suitable in these situations.

When n equals infinity, a present sum P will accrue interest of Pi for every future interest period. For the principal sum P to continue undiminished (an essential requirement for n equal to infinity), the end-of-period sum A that can be disbursed is Pi.

$$P \rightarrow P + Pi \rightarrow P + Pi \rightarrow P + Pi \rightarrow \cdots$$
$$\searrow A \qquad \searrow A \qquad \searrow A$$

When $n = \infty$ the fundamental relationship between P, A, and i is:

$$A = Pi$$

Some form of this equation is used whenever there is a problem with an infinite analysis period.

ANNUAL COST

The annual cost method is more accurately described as the method of Equivalent Uniform Annual Cost (EUAC) or, where the computation is of benefits, the method of Equivalent Uniform Annual Benefit (EUAB).

Criteria

For each of the three possible categories of problems there are annual cost criteria for economic efficiency.

Category	Annual Cost Criterion
Fixed Input	Maximize the equivalent uniform annual benefits, that is, Maximize EUAB.
Fixed Output	Minimize the equivalent uniform annual cost, that is, Minimize EUAC.
Neither Input nor Output fixed	Maximize [EUAB - EUAC]

Application of the Annual Cost Method

In the present worth section we pointed out that the present worth method requires that there be a common analysis period for all alternatives. This same restriction does not apply in all annual cost calculations, but it is important to understand the circumstances that justify comparing alternatives with different service lives.

It is frequently the case that an analysis is to provide for a more or less continuing requirement. One might need to pump water from a well, for example, as a continuing requirement. Regardless of whether the pump has a useful service life of 6 years or 12 years, we would select the one whose annual cost was a minimum. And this would still be the case if the pump useful lives were the more troublesome 7 and 11 years, respectively. Thus if we can assume a continuing need for an item, an annual cost comparison among alternatives of different service lives is valid.

The underlying assumption made in these situations is that when the shorter lived alternative has reached the end of its useful service life, it can be replaced with an identical item with identical costs, etc. Therefore the EUAC of the initial alternative for a relatively short period of time is equal to the EUAC for the continuing series of replacements.

If, on the other hand, there was a specific requirement in some situation to pump water for 10 years, then each pump must be evaluated to see what costs would be incurred during the analysis period and what salvage value, if any, could be recovered at the end of the analysis period. The annual cost analysis would need to consider the actual circumstances of the situation.

In the E. E. professional examination the problems frequently are most readily solved by the annual cost method. And the underlying "continuing requirement" is usually present, so that an annual cost comparison of unequal lived alternatives is an appropriate method of analysis.

RATE OF RETURN

In the rate of return method the typical situation is where there is a cash flow representing both costs and benefits. The rate of return may be defined as the interest rate when

PW of Cost = PW of Benefit

or EUAC = EUAB.

These calculations frequently require trial and error solution.

Criteria

Category	Rate of Return Criterion
Fixed Input	Maximize the rate of return.
Fixed Output	Maximize the rate of return.
Neither Input nor Output fixed	TWO ALTERNATIVES Compute the incremental rate of return on the cash flow representing the differences between the alternatives. If this rate of return is ⩾ minimum attractive rate of return (MARR), choose the higher cost alternative; otherwise choose the lower cost alternative. THREE OR MORE ALTERNATIVES Incremental analysis is required. The details are explained in the following section.

Rate of Return When Neither Input Nor Output Fixed

The rate of return method becomes complicated in the general category of neither input nor output fixed.

When there is a situation of two mutually exclusive (doing one precludes doing the other) alternatives we know we must select one of the two alternatives, but we cannot select them both. On Page 129, two alternatives were described in terms of cash flows.

Year	Alternative A	B
0	-2000	-2800
1	+800	+1100
2	+800	+1100
3	+800	+1100

If one considers 5% the minimum attractive rate of return (MARR), which alternative should be selected? A computation of the rate of return for each alternative would show that both alternatives have a rate of return in excess of 5% (Alt. A = 9.7%, Alt. B = 8.7%).

A conventional assumption in economic analysis is that we will select the larger investment (rather than the smaller one) if the rate of return on the additional investment is greater than or equal to the minimum attractive rate of return (MARR). This means there will be situations where the alternative selected is not the alternative with the largest rate of return.

To decide which alternative in the example to select, compute the cash flow that represents the difference between the alternatives.

Year	Alternative A	B	Difference Between Alternatives B-A
0	-2000	-2800	-800
1	+800	+1100	+300
2	+800	+1100	+300
3	+800	+1100	+300
Rate of Return	9.7%	8.7%	6.1%

Since the rate of return on the difference between the alternatives (6.1%) exceeds the 5% MARR, the increment of additional investment is desirable. In this example we would select Alternative B.

When there are three or more mutually exclusive alternatives one must proceed following the same general logic presented for two alternatives. The components of incremental analysis are:

1. Compute the rate of return for each alternative. Reject any alternative where the rate of return is less than the given MARR. [This step is not essential but helps to

immediately eliminate unacceptable alternatives. One must insure, however, that the lowest cost alternative has a rate of return $\geq$ MARR.]

2. Rank the remaining alternatives in their order of increasing cost.
3. Examine the difference between the two lowest cost alternatives as described for the two alternative problem. Select the better of the two alternatives and reject the other one.
4. Take the preferred alternative from Step 3. Add the next higher alternative and proceed with another two alternative comparison.
5. Continue until all alternatives have been examined and the best of the multiple alternatives has been identified.

Application of Rate of Return Analysis

One advantage of the rate of return method is that it gives a result that is readily understood. Another is that it is very straightforward to compute in the two categories of Fixed Input and Fixed Output.

It is in the general category of neither input nor output fixed where rate of return becomes cumbersome. In this situation it is frequently faster to solve a problem by other techniques. Also, there may be uncertainty on how to compute the difference between two alternatives when they have different useful lives. This will, of course, depend on the facts in the particular situation being examined.

DEPRECIATION

Depreciation of capitalized equipment is an important component of many after-tax economic analyses. For this reason one must understand the fundamentals of depreciation accounting.

Depreciation is defined, in its accounting sense, as the systematic allocation of the cost of a capital asset over its useful life. In computing a schedule of depreciation charges three items are considered.

1. Cost the property, P.
2. Useful life in years, n.
3. Salvage value of the property at the end of its useful life, F.

Three principal methods of depreciation are:

Straight Line Depreciation

$$\text{Depreciation charge in any year} = \frac{P - F}{n}$$

Sum-Of-Years-Digits Depreciation

$$\text{Depreciation charge in any year} = \frac{\text{Remaining Useful Life at Beginning of Year}}{\text{Sum of Years Digits for Total Useful Life}} (P - F)$$

where Sum-Of-Years-Digits = $1+2+3+\cdots+n = \frac{n}{2}(n + 1)$

Double Declining Balance Depreciation

$$\text{Depreciation charge in any year} = \frac{2}{n} (P - \text{Depreciation charges to date})$$

INCOME TAXES

Income taxes represent another of the various kinds of disbursements encountered in an economic analysis. The starting point in any after-tax computation is the before-tax cash flow.

Generally, the before-tax cash flow contains three types of entries.

1. Disbursements of money to purchase capital assets. These expenditures create no direct tax consequence for they are the exchange of one asset (cash) for another (capital assets, like equipment).
2. Periodic receipts and/or disbursements representing operating income and/or expenses. These increase or decrease the year-by-year tax liability of the firm.
3. Receipt of money from the sale of capital assets, usually in the form of salvage value when the equipment is removed. The tax consequence depends on the relationship between the book value (cost - depreciation taken) of the asset and its salvage value.

Situation	Tax Consequence
Salvage value > Book value	Capital gain on difference
Salvage value = Book value	No tax consequence
Salvage value < Book value	Capital loss on difference

After the before-tax cash flow, the next step is to compute the depreciation schedule for any capital assets. Taxable income is the taxable component of the before-tax cash flow minus the depreciation. The income tax is the taxable income times the appropriate tax rate. Finally, the after-tax cash flow is the before-tax cash flow adjusted for income taxes.

To organize these data it is customary to arrange them in the form of a cash flow table as follows.

Year	Before Tax Cash Flow	Depreciation	Taxable Income	Income Taxes	After Tax Cash Flow
0	.		.	.	.
1	.	.	.	.	.
.	.	.	.	.	.

REFERENCES

Grant & Ireson: Principles of Engineering Economy, 5th edition. New York: Ronald Press, 1970.

Thuesen, Fabrycky & Thuesen: Engineering Economy, 4th edition. Englewood Cliffs, New Jersey: Prentice-Hall, Inc., 1971.

PROBLEMS AND SOLUTIONS

ECONOMICS 1.

An engineer is faced with the prospect of a fluctuating future budget for the maintenance of a particular machine. During each of the first five years $1000 per year will be budgeted. During the second five years the annual budget will be $1500 per year. In addition, $3500 will be budgeted for an overhaul of the machine at the end of the fourth year, and another $3500 for an overhaul at the end of the eighth year.

The engineer wonders what uniform annual expenditure would be equivalent to these fluctuating amounts, assuming compound interest at 6% per annum. Compute the equivalent uniform annual expenditure for the 10 year period.

Solution

A diagram of the projected budget amounts is as follows:

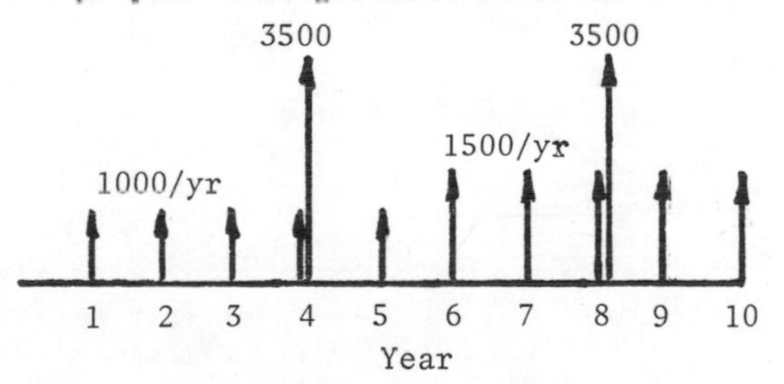

Compute the present worth of the various future amounts. Then compute the equivalent uniform annual amount.

$$\begin{aligned}PW &= 1000(P/A,6\%,5) + 3500(P/F,6\%,4) + 1500(P/A,6\%,5)(P/F,6\%,5) \\ &\quad + 3500(P/F,6\%,8) \\ &= 1000(4.212) + 3500(0.7921) + 1500(4.212)(0.7473) \\ &\quad + 3500(0.6274) \\ &= 4212 + 2772 + 4721 + 2196 = \$13,901\end{aligned}$$

$$\begin{aligned}\text{Equivalent uniform annual amount} &= 13,901(A/P,6\%,10) \\ &= 13,901(0.1359) = \$1889\end{aligned}$$

ECONOMICS 2.

A new office building was constructed five years ago by a consulting engineering firm. At that time the firm obtained a bank loan for \$100,000 with a 6% annual interest rate, compounded quarterly. The terms of the loan called for equal quarterly payments for a 10 year period with the right of prepayment at any time without penalty.

Due to internal changes in the firm, it is now proposed to refinance the loan through an insurance company. The new loan is planned for a 20 year term with an interest rate of 8% per annum, compounded quarterly. The insurance company has a one time 5% service charge. This new loan also calls for equal quarterly payments.

REQUIRED:

(a) What is the balance due on the original mortgage (principal) if all payments have been made through a full five years?

(b) What will be the difference between the equal quarterly payments in the existing arrangement and the revised proposal?

Solution

(a)

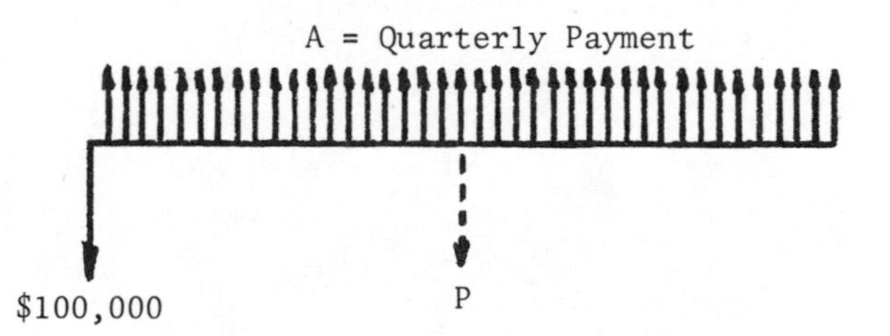

6% per year, compounded quarterly for 10 years

Therefore i = 1-1/2% per interest period
n = 40 interest periods

A = 100,000(A/P,1-1/2%,40) = 100,000(0.0334) = \$3340

P = 3340(P/A,1-1/2%,20) = 3340(17.169) = \$57,344

The balance due is \$57,344

(b)

Service charge = 0.05P

Amount of New Loan = 1.05P = 1.05(57,344) = \$60,211

Quarterly payments on new loan = 60,211(A/P,2%,80)
= 60,211(0.0252) = \$1517

Difference in quarterly payments between existing loan and new loan is (\$3340 - \$1517) = \$1823

ECONOMICS 3.

An investor is considering buying a 20-year corporate bond. The bond has a face value of $1000 and pays 6% interest per year in two semiannual payments. Thus the purchaser of the bond would receive $30 every 6 months and in addition he would receive $1000 at the end of 20 years, along with the last $30 interest payment.

If the investor thought he should receive 8% interest, compounded semiannually, how much would he be willing to pay for the bond?

Solution

The investor would be willing to pay the present worth of the future benefits, computed at 8% compounded semiannually.

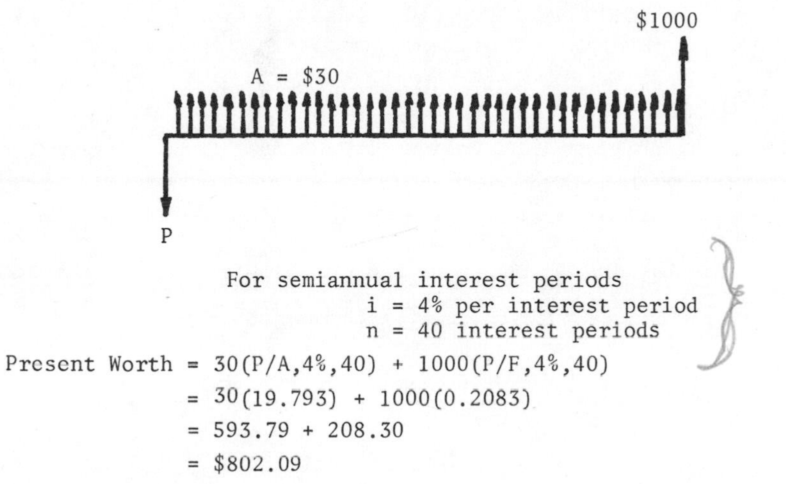

For semiannual interest periods

$i = 4\%$ per interest period

$n = 40$ interest periods

$$\begin{aligned}\text{Present Worth} &= 30(P/A,4\%,40) + 1000(P/F,4\%,40)\\ &= 30(19.793) + 1000(0.2083)\\ &= 593.79 + 208.30\\ &= \$802.09\end{aligned}$$

ECONOMICS 4.

Rental equipment is for sale at $110,000. A prospective buyer estimates he would hold the equipment for 12 years, during which time the average annual disbursements for all purposes in connection with its ownership and operation would be $6000.

The prospective buyer believes the equipment could be sold for $80,000 after sales expenses at the end of 12 years. He sets the minimum attractive return on this type of investment to be 7% before income taxes. Estimated average annual receipts from equipment rentals are $14,400.

On the basis of these estimates, what is the maximum price he should pay for the equipment?

Solution

Compute the cash flow.
Let X = maximum purchase price.

Year	Before Tax Cash Flow
0	-X
1	+14,400 - 6,000
2	↓
3	↓
.	↓
.	↓
.	↓
12	{ +80,000

Maximum purchase price = present worth of future benefits

$$X = (14{,}400 - 6000)(P/A,7\%,12) + 80{,}000(P/F,7\%,12)$$
$$= 8400(7.943) + 80{,}000(0.444)$$
$$= 66{,}700 + 35{,}500 = \underline{\underline{\$102{,}200}}$$

ECONOMICS 5.

Imagine that you must take one of two courses of action in the future which will cost as follows:

Year	Alternative A	Alternative B
0	$1300	$ 0
1		100
2		200
3		300
4		400
5		500
Total	$1300	$1500

Using present worth of the two alternatives with interest equal to 6%, which is the more economical?

Solution

Alternative A

PW of Cost = $1300

Alternative B

$$\text{PW of Cost} = 100(P/F,6\%,1) + 200(P/F,6\%,2) + 300(P/F,6\%,3) + 400(P/F,6\%,4) + 500(P/F,6\%,5)$$
$$= 100(0.943) + 200(0.890) + 300(0.840) + 400(0.792) + 500(0.747)$$
$$= 94 + 178 + 252 + 317 + 373$$
$$= \$1214$$

Choose the alternative with the least PW of Cost. Choose B.

ECONOMICS 6.

A tunnel excavation project is located 5 miles from the nearest electric power source. The work schedule for the job will be 24 hours per day, 6 days per week. The total time for completion of the project is three years. The power demand for the first two years of construction will be 4500 KW. The power demand for the final year will be 3000 KW.

The local electric utility has agreed to provide power lines and substations for a 4500 KW demand at a cost of \$135,000. Of this cost, \$68,000 is for the erection and removal of the equipment. The salvage value of the power lines and substations at the end of 3 years will be 60% of their original cost. The utility's charge for power is \$0.008 per KWH.

An alternative to using utility power is for the constructor to supply his own power needs with diesel generators. An equipment manufacturer has quoted a price of \$60,000 each for 750 KW generators. These generators are powered by 1300 hp diesel engines which consume 0.04 gallons of diesel fuel per hp per hour. The generators have a useful life of ten years. Depreciation will be on a straight line basis. The generators probably can be sold at their depreciated value. Assume that all payments for power to the utility, and all payments for diesel fuel are made at year end. Diesel fuel costs \$0.15 per gallon. Assume a 7% annual interest rate.

REQUIRED:

(a) Prepare a cost analysis for the two alternatives showing all calculations.

(b) Which of the alternatives would be the more economical to the constructor?

Solution

Alternative I - Power from Utility Company

Cost of Installation = \$135,000

Terminal Value at end of 3 years = 0.6(135,000 - 68,000)
= \$40,200

Annual Power Charge - First Two Years:

$$24 \frac{\text{hrs}}{\text{day}} \times 6 \frac{\text{days}}{\text{week}} \times 52 \frac{\text{weeks}}{\text{year}} \times 4500 \text{ KW} \times \$0.008/\text{KWH} = \$269,600$$

Power Charge - Third Year:

$$\frac{3000}{4500}(269,600) = \$179,700$$

Alternative II - Diesel Generators

Cost of six Generators = 6 x 60,000 = \$360,000

Disposal Plan: Remove two Generators at the end of the second year and the remaining four Generators at the end of the third year.

Terminal value of first two Generators:
0.8(120,000) = \$96,000 at the end of the second year

Terminal value of the last four Generators:
0.7(240,000) = \$168,000 at the end of the third year

Hourly Fuel Cost = 6 x 1300 hp x 0.04 x 0.15 = \$46.80 per hour

Annual Fuel Charge during the first two years

$$24 \frac{\text{hrs}}{\text{day}} \times 6 \frac{\text{days}}{\text{week}} \times 52 \frac{\text{weeks}}{\text{year}} \times \$46.80 \text{ per hour} = \$350,400$$

Annual Fuel Charge for the third year

$$\frac{3000}{4500}(350,400) = \$233,600$$

From these data we can write the cash flow for each alternative:

Year	Alternative I Power From Utility	Alternative II Diesel Generators
0	-135,000	-360,000
1	-269,600	-350,400
2	-269,600	-350,400 +96,000
3	-179,700 +40,200	-233,600 +168,000

Present Worth Analysis

Alternative I - Power from Utility Company

Present Worth of Cost = 135,000 + 269,600(P/A,7%,2) + (179,700 - 40,200)(P/F,7%,3)
= 135,000 + 269,600(1.808) + 139,500(0.8163)
= 135,000 + 487,400 + 113,900
= 736,300

Alternative II - Diesel Generators

Present Worth of Cost = 360,000 + 350,400(P/A,7%,2) -96,000(P/F,7%,2) + 65,600(P/F,7%,3)
= 360,000 + 350,400(1.808) -96,000(0.8734) + 65,600(0.8163)
= 963,200

(b) Decision

Choose the alternative with the smaller Present Worth of Cost. Buy power from the utility company.

ECONOMICS 7.

You are the engineer for a large ranch. A three year old deep well pump has decreased considerably in flow because of wear. It now delivers 1200 gallons per minute from a pumping level of 140 feet against open discharge. Power used is 55 kilowatts. Although the flow is adequate for the need of 800 acre-feet per year, it has been suggested that good economics demands the pump be repaired.

For $5250 it could be repaired to its original condition of 72% overall efficiency. The ranch accountant tells you money is worth 7% and that the accounting life of such equipment is 10 years, with zero salvage value. Power costs $0.02 per kilowatt hour. Should you order the pump to be repaired?

Solution

$$\text{Hours of pumping/year} = \frac{800 \text{ af} \times 43{,}560 \frac{\text{cf}}{\text{af}} \times 7.481 \frac{\text{gal}}{\text{cf}}}{1200 \frac{\text{gal}}{\text{min}} \times 60 \frac{\text{min}}{\text{hr}}} = 3620.8 \text{ hrs}$$

Existing Pump

$$\text{Power Cost/year} = 55 \text{ kw} \times 3620.8 \frac{\text{hrs}}{\text{yr}} \times 0.02 \frac{\$}{\text{kw-hr}} = \$3982.88$$

Repaired Pump

At 1200 gpm

$$\text{KW input} = \frac{1200 \frac{\text{gal}}{\text{min}} \times 8.33 \frac{\text{lb}}{\text{gal}} \times 140 \text{ ft}}{0.72 \text{ eff} \times 33{,}000 \frac{\text{ft-lb/min}}{\text{hp}} \times \frac{1}{0.746} \frac{\text{hp}}{\text{KW}}}$$

$$= 43.94 \text{ KW}$$

Power Cost/year = 43.94 x 3620.8 x $0.02 = $3181.96

Annual Power Saving with Repaired Pump = $3982.88 - $3181.96
= $800.92

Present Worth of Power Saving over the next seven years
= $800.92(P/A,7%,7) = $800.92(5.389) = $4316.16

Before making a decision, a more detailed analysis of the pump wear may be needed. It is not clear that the pump will continue at this performance level or that the repair will insure an overall 72% efficiency for the next 7 years. Both of these are important assumptions in the analytical solution. Based on the calculations, however, the Present Worth of Benefits is less than the \$5250 cost, indicating the repairs should not be made.

ECONOMICS 8.

What present investment is necessary to secure a perpetual income of \$1200 a year if interest is 4% per annum?

Solution

$$\text{Capitalized cost } P = \frac{A}{i} = \frac{1200}{0.04} = \$30,000$$

ECONOMICS 9.

A trust fund is to be established to

(a) Provide \$750,000 for the construction and \$250,000 for the initial equipment of an Electrical Engineering Laboratory.

(b) Pay the laboratory annual operating costs of \$150,000 per year.

(c) Pay for \$100,000 of replacement equipment every four years, beginning four years from now.

At 4% interest, how much money would be required in the trust fund to provide for the laboratory and equipment and its perpetual operation and equipment replacement?

Solution

This problem is composed of three components.

Required money in trust fund = \$1,000,000 initial construction and equipment money
+ money for annual operation
+ money for periodic equipment replacement.

For $n = \infty$

$$A = Pi \quad \text{or } P = \frac{A}{i}$$

For (a) $P = \$1,000,000$

(b) $$P = \frac{A}{i} = \frac{\$150,000}{0.04} = \$3,750,000$$

(c) This situation is more complex. We do not know what equivalent amount A will provide for the necessary replacement equipment.

The perpetual series:

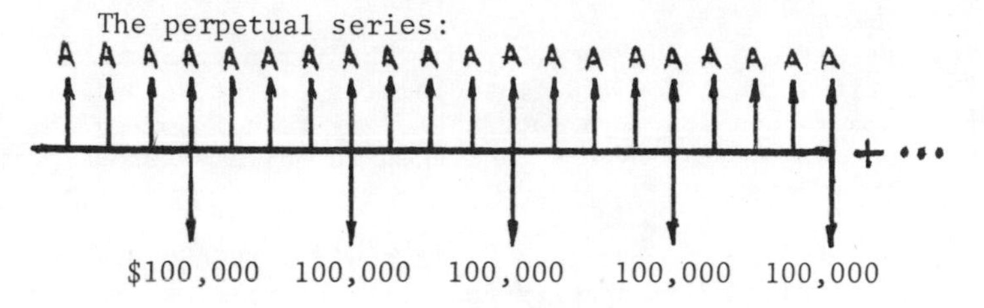

We can solve one portion of the perpetual series for A:

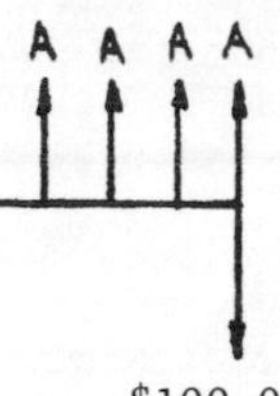

$A = \$100{,}000(A/F,4\%,4) = \$100{,}000(0.2355) = \$23{,}550$

This value of A computed for the four year period is the same as the value of A for the perpetual series.

For the perpetual series:

$$P = \frac{A}{i} = \frac{\$23{,}550}{0.04} = \$589{,}000$$

Required money in trust fund = \$1,000,000 + 3,750,000 + 589,000
= <u>\$5,339,000.</u>

ECONOMICS 10.

A nearby city has developed a plan which will provide for future municipal water needs. The plan proposes an aqueduct which passes through 1000 feet of tunnel in a nearby mountain.

Two alternatives are being considered. The first proposes to build a full-capacity tunnel now (10 foot diameter). The second proposes to build a half-capacity (7.5 foot diameter) tunnel now which should be adequate for 20 years, and then to build a second parallel half-capacity tunnel.

The full-capacity tunnel A can be built now for $556,000 as compared to $402,000 for one half-capacity tunnel. The costs of repair to the tunnel lining for the full-capacity tunnel are estimated to be $0.64 per square foot of tunnel every 10 years. For the half-capacity tunnel B the cost is estimated as $0.70 per square foot every 10 years.

The friction losses in the half-capacity tunnel will be somewhat greater, and it is estimated that the pumping costs in the single half-capacity line will increase by $2000 per year, and by $4000 per year after the second half-capacity tunnel has been placed in service.

REQUIRED: Using an annual interest rate of 7%, what is the total capitalized cost for each of the two alternatives?

Solution

Tunnel Lining

Full Capacity Tunnel

Surface Area = $10\pi(1000 \text{ feet}) = 31{,}416 \text{ ft}^2$

Repair Cost = 31,416($0.64) = $20,106

Half Capacity Tunnel

Surface Area = $7.5\pi(1000 \text{ feet}) = 23{,}562 \text{ ft}^2$

Repair Cost = 23,562($0.70) = $16,493 per tunnel

Alternative A - Full Capacity Tunnel

$$\begin{aligned} & \qquad\quad \text{Initial Cost} \qquad \text{Lining Repair} \\ \text{Capitalized Cost} &= \$556{,}000 + \frac{\$20{,}106(A/F,7\%,10)}{0.07} \\ &= \$556{,}000 + \frac{\$20{,}106(0.0724)}{0.07} \\ &= \$556{,}000 + \$20{,}800 = \underline{\$576{,}800} \end{aligned}$$

Alternative B - Half Capacity Tunnels

First Half Capacity Tunnel

$$\text{Capitalized Cost} = \underset{\text{Initial Cost}}{\$402{,}000} + \underset{\text{Lining Repair}}{\frac{16{,}493(A/F,7\%,10)}{0.07}} + \underset{\text{Extra Pumping}}{\frac{2000}{0.07}}$$

$$= \$402{,}000 + \frac{16{,}493(0.0724)}{0.07} + \frac{2000}{0.07}$$

$$= \$447{,}600$$

Second Half Capacity Tunnel

The capitalized cost of the second half-capacity tunnel 20 years hence is equal to the present capitalized cost of the first half-capacity tunnel.

$$\text{The present capitalized cost} = \frac{\text{capitalized cost}}{\text{20 years hence}} \times (P/F,7\%,20)$$

$$= \$447{,}600(0.2584)$$

$$= \$115{,}700$$

Capitalized Cost for the half-capacity tunnels = 447,600 + 115,700
= $563,300

ECONOMICS 11.

A pumping service is required for ten years at a remote location, and gasoline and electric power are being considered on the basis of the following data.

	Gasoline	Electric
First Cost	\$1200	\$3000
Life in Years	5 yrs	10 yrs
Salvage Value	\$150	\$300
Annual Operating Cost	\$600	\$375
Annual Repairs	\$150	\$75
Annual Taxes (% of first cost)	1%	1%

Analyze the two installations on the basis of annual costs using 6 percent interest to determine which is the least expensive.

Solution

Equivalent Uniform Annual Cost (EUAC)

$$= (P-F)(A/P,i\%,n) + Fi + \text{Annual Costs}$$

Gasoline

$$\text{EUAC} = (1200 - 150)(A/P,6\%,5) + 150(0.06) + 600 + 150 + 0.01(1200)$$

$$= 1050(0.2374) + 9 + 600 + 150 + 12 = \$1020.27$$

Electric

$$\text{EUAC} = (3000 - 300)(A/P,6\%,10) + 300(0.06) + 375 + 75 + 0.01(3000)$$

$$= 2700(0.1359) + 18 + 375 + 75 + 30 = \$864.93$$

The electric driven pump has the smaller annual cost.

ECONOMICS 12.

A snow loading machine costing \$30,000 requires four operators at \$36 per day. The machine can do the work of 50 hand shovelers at \$21 per day. Fuel, oil and maintenance for the machine amount to \$60 per day. Assume the life of the machine is 8 years with no salvage value. If interest on money is 6%, how many days of snow removal per year are necessary to make purchase of the machine economical?

Solution

Let Annual Cost (EUAC) of hand loading = Annual Cost (EUAC) of machine loading, and solve for X, equal to days of snow removal per year.

$$50(\$21)(X) = 4(\$36)(X) + \$60(X) + \$30,000(A/P,6\%,8)$$

$$1050\,X = 144\,X + 60\,X + 30,000(0.161)$$

$$X = 5.7 \text{ days/year}$$

ECONOMICS 13.

A machine part, operating in a corrosive atmosphere, is made of low-carbon steel, costs $350 installed and lasts six years. If the part is treated for corrosion resistance it will cost $700 installed. How long must the treated part last to be a better investment if money is worth 7%?

Solution

The annual cost of the untreated part is

$$\$350(A/P,7\%,6) = 350(0.2098) = \$73.43$$

The annual cost of the treated part must be at least this low, so

$$\$73.43 = \$700(A/P,7\%,n)$$

$$(A/P,7\%,n) = \frac{73.43}{700} = 0.1049$$

From compound interest tables:

n	(A/P,7%,n)
16 yrs	0.1059
17 yrs	0.1024

By linear interpolation we get

$$n = 16 + (1)\frac{0.1059 - 0.1049}{0.1059 - 0.1024} = 16 + \frac{0.0010}{0.0035} = 16.3 \text{ years}$$

Thus the treated part must last longer than 16.3 years to be a better investment than the untreated part.

ECONOMICS 14.

An engineer must select a pump for service requiring 14,000 gallons per minute at 11 feet total dynamic head. The specific gravity of the liquid is 1.45. Supplier A offers a pump of all-iron construction with an efficiency of 70% at a price of $5600; Supplier B offers a pump of all-monel construction with an efficiency of 75% at a price of $14,700. Iron contamination of the product can be tolerated.

Installation costs and/or replacement costs for either pump are $1000. Routine maintenance costs are assumed the same in either case, and power is available at $0.01 per kilowatt-hour. Operation will be 24 hours per day for 300 days per year.

It is expected that the monel pump will last for 20 years, at which time the plant will be obsolete and the monel pump will have a salvage value of 25% of its original cost.

If money is available at 6%, what minimum service life must be expected from the all-iron pump to justify its selection assuming no salvage value?

Solution

Input Power at 100% efficiency

$$14{,}000\ \frac{\text{gal}}{\text{min}} \times (8.33 \times 1.45)\frac{\text{lbs}}{\text{gal}} \times 11\ \text{ft} \times \frac{1\ \text{hp}}{33{,}000\frac{\text{ft-lb}}{\text{min}}} \times 0.746\ \frac{\text{KW}}{\text{hp}}$$

Input Power = 42 Kilowatts

Power Cost

All-Iron Pump: $42\ \text{KW} \times \frac{1}{0.70\ \text{eff}} \times 24\ \frac{\text{hrs}}{\text{day}} \times 300\ \frac{\text{days}}{\text{yr}} \times 0.01\ \frac{\$}{\text{Kwh}}$

= $4320 per year

Monel Pump: $42 \times \frac{1}{0.75} \times 24 \times 300 \times \$0.01 = \$4032$ per year

Pump Costs	Initial Cost	Installation	Salvage Value
All-Iron Pump	$ 5,600	$1,000	0
Monel Pump	14,700	1,000	0.25(14,700) = 3675

Compute the minimum service life of the iron pump (life = n years) such that

$$\text{Annual Cost}_{\text{iron pump}} = \text{Annual Cost}_{\text{monel pump}}$$

$$\begin{aligned}\text{Annual Cost}_{\text{monel pump}} &= (14{,}700 + 1{,}000)(A/P,6\%,20) \\ &\quad - 3675(A/F,6\%,20) + 4032 \\ &= 15{,}700(0.0872) + 4032 - 3675(0.0272) \\ &= 1369 + 4032 - 100 \\ &= \$5301\end{aligned}$$

$$\text{Annual Cost}_{\text{iron pump}} = (5600 + 1000)(A/P,6\%,n) + 4320$$

Equating the two annual costs:

$$\$5301 = (6600)(A/P,6\%,n) + 4320$$

$$(A/P,6\%,n) = \frac{5301 - 4320}{6600} = 0.1486$$

From 6% interest table: n slightly less than 9 years

ECONOMICS 15.

GIVEN: The Highridge Water District needs an additional supply of water from Steep Creek. The engineer has selected two plans for comparison:

(A) GRAVITY PLAN: Divert water at a point 10 miles up Steep Creek and carry it through a pipeline by gravity to the District.

(B) PUMPING PLAN: Divert water at a point near the District and pump it through 2 miles of pipeline to the District. The pumping plant can be built in two stages, with one-half capacity installed initially and the other one-half 10 years later.

All costs are to be repaid within 40 years, with interest at 5%. Salvage values can be ignored. During the first 10 years, the average use of water will be less than during the remaining 30 years. Costs are as follows:

	Gravity	Pumping
Initial Investment	$2,800,000	$1,400,000
Additional Investment in 10th year	none	200,000
Operation, Maintenance and Replacements	10,000/yr	25,000/yr
Power Cost		
Average first 10 years	none	50,000/yr
Average next 30 years	none	100,000/yr

REQUIRED: Select the more economical plan. Show your computations.

Solution

Equivalent Uniform Annual Cost Comparison

Gravity Plan

Initial Investment 2,800,000(A/P,5%,40) = 2,800,000(0.0583)	163,240
Operation, Maintenance & Replacements	10,000
Annual Cost =	$173,240

Pumping Plan

Initial Investment 1,400,000(A/P,5%,40) = 1,400,000(0.0583)	81,620
Additional Investment in 10th year 200,000(P/F,5%,10)(A/P,5%,40) 200,000(0.6139)(0.0583)	7,160
Operation, Maintenance & Replacements	25,000
Power Cost 50,000/year for 40 years	50,000
Additional 50,000/year for last 30 years 50,000(F/A,5%,30)(A/F,5%,40) 50,000(66.439)(0.0083)	27,570
Annual Cost =	$191,350

Choose the alternative with the smaller equivalent uniform annual cost. Choose the Gravity Plan.

Alternate Method Computation:

Present Worth Comparison

Gravity Plan

Initial Investment	2,800,000
Operation, Maintenance & Replacements 10,000(P/A,5%,40) = 10,000(17.159)	171,590
Present Worth =	$2,971,590

Pumping Plan

Initial Investment	1,400,000
Additional Investment in 10th year 200,000(P/F,5%,10) = 200,000(0.6139)	122,780
Operation, Maintenance & Replacements 25,000(P/A,5%,40) = 25,000(17.159)	428,980
Power First 50,000/year 50,000(P/A,5%,40) = 50,000(17.159)	857,950
Additional 50,000/year for last 30 years 50,000(P/A,5%,30)(P/F,5%,10) 50,000(15.372)(0.6139)	471,840
Present Worth =	$3,281,550

Choose the alternative with the smaller present worth of cost. Choose the Gravity Plan.

ECONOMICS 16.

Five years ago a dam was constructed on a stream for the purpose of impounding irrigation water. It was also expected to provide flood protection for the area below the dam. Last winter a one-hundred-year flood caused extensive damage both to the dam and to the surrounding area. This may not have been surprising in view of the fact that the dam was designed for a fifty-year flood.

It is estimated that the cost to repair the dam now will be $250,000. Damage in the valley below probably amounts to $750,000. If the spillway is redesigned and certain other improvements are made at a total cost of $500,000 (including repair), the dam may be expected to withstand a one-hundred-year flood without sustaining damage. However, since the storage capacity will not be increased, the probability of damage to the valley will be unchanged. Additional storage would be required to eliminate concern for possible future damage. A second dam can be constructed up the river from the existing dam for $1,000,000. The capacity of the second dam would be more than adequate to provide the desired protection. If the second dam is built, redesign of the existing dam will not be necessary. Repairs to the existing dam will be required regardless.

The valley development is expected to be complete in ten years. A new flood in the meantime may average a $1,000,000 loss, after that time you assume the probable cost will remain at $2,000,000. This is an obvious oversimplification, but is felt to be a reasonable approximation.

In reviewing the decision that must be made now, the investigating board is weighing the recommendation of another consulting engineer who recommends no other action than the repair of the spillway. His argument is that another such flood would not occur for another one hundred years. He concedes that a fifty-year flood is also likely to cause damage, but the spillway would be adequate and the total damage would be about $200,000. Likewise, a twenty-five-year flood would also cause damage amounting to about $50,000.

The consulting engineer also reminds the review board that the community is now taxed to repay a twenty-year serial bond issue of $5,000,000 at 5 percent interest, floated five years ago to finance the original construction. The dam was "sold" to the people then as being good for a fifty-year life, and they are understandably angry over the present situation. The main structure is still good, and its life expectancy now is still fifty years.

REQUIRED:

Give your economic evaluation of the situation. Support your argument with an economic analysis based on an equivalent annual cost comparison of the three alternatives suggested. There may be other alternatives which have not been explored in the above discussion, but you are not expected to extend the problem by considering such, nor are you expected to discuss political or irreducible factors. Point out any errors in the other engineer's reasoning.

Solution

	Flood	Probability of Damage in any year = 1/year	Downstream Damage	Spillway Damage
	25 year	0.04	50,000	-
	50 year	0.02	200,000	-
In next 10 years	100 year	0.01	1,000,000	250,000
Thereafter	100 year	0.01	2,000,000	250,000

Note that we have assumed the $1 million and $2 million in projected damages do not include spillway damage.

Three Alternatives

1. Repair the existing dam ($250,000).
   Make no alterations.

2. Repair the dam and redesign the spillway to carry a 100-year flood ($500,000).

3. Repair the existing dam ($250,000).
   Build a flood control dam upstream ($1,000,000).

Equivalent Uniform Annual Cost

In economic analysis we need be interested only in the differences between the alternatives. Since at least $250,000 worth of work will be done now on the existing dam in all alternatives, this $250,000 may either be included or ignored. Here it will be left out of the analysis.

Alternative 1.

Spillway Damage

The probability that the spillway capacity will be equaled or exceeded in any year is 0.02.

Damage if spillway capacity is exceeded = $250,000.

Expected annual cost of spillway damage
= Damage(Probability of its occurrence)
= 250,000(0.02) = $5000.

Alternative 1 - Continued

Downstream Damage During Next 10 Years

Flood	Probability that flow will be equaled or exceeded*	Damage	Incremental damage over more frequent flood	Annual Cost of flood risk
25 yr	0.04	50,000	50,000	2000
50 yr	0.02	200,000	150,000	3000
100 yr	0.01	1,000,000	800,000	8000

Next 10-year expected annual cost of downstream damage: $13,000

*An N-year flood will be equaled or exceeded at an average interval of N years.

Downstream Damage After 10 Years

Following the same logic as above,

Expected annual cost of downstream damage
= 2000 + 3000 + 0.01(2,000,000 - 200,000) = $23,000

Present Worth of Cost of Expected Spillway Damage plus Expected Downstream Damage

PW = 5000(P/A,7%,50) + 13,000(P/A,7%,10)
+ 23,000(P/A,7%,40)(P/F,7%,10)

= 5000(13.801) + 13,000(7.024) + 23,000(13.332)(0.5083)
= 69,005 + 91,312 + 155,863
= 316,180

Equivalent Annual Cost = 316,180(A/P,7%,50) = 316,180(0.0725)
= $22,920.

Alternative 2.

(Repair the dam and redesign the spillway.)

Additional cost to redesign and reconstruct the spillway = 250,000

Downstream Damage - Same as Alternative 1.

Present Worth of Cost to Reconstruct Spillway plus Expected Downstream Damage

PW = 250,000 + 13,000(P/A,7%,10)
+ 23,000(P/A,7%,40)(P/F,7%,10)

= 250,000 + 13,000(7.024) + 23,000(13.332)(0.5083)
= 250,000 + 91,312 + 155,863
= 497,175

Equivalent Annual Cost = 497,175(A/P,7%,50) = 497,175(0.0725)
= $36,050.

Alternative 3.

(Repair the dam and build flood control dam upstream.)

Cost of flood control dam = 1,000,000

Equivalent Annual Cost = 1,000,000(A/P,7%,50)
= 1,000,000(0.0725)
= $72,500.

Since we are dealing under conditions of risk, it is not possible to make an absolute statement concerning which alternative will result in the least cost to the community. Using a probabilistic approach, however, we can select the alternate which is most likely to result in the least equivalent cost. From the Equivalent Uniform Annual Cost (EUAC) calculations we see that Alternative 1 (Do nothing but repair the damage.) is most likely to result in the least equivalent cost to the community.

One must be careful not to confuse the frequency of a flood and when it might be expected to occur. The occurrence of a 100-year flood this year is no guarantee that it won't happen again next year. In any 50 year period, for example, there are 4 chances in 10 that a 100-year flood (or greater) will occur.

It was suggested by a consulting engineer that no alterations be made now because the project was originally designed to last fifty years. This reasoning is not sound. In the five years since the original construction a number of changes may have taken place that were not anticipated in the earlier planning. More rapid development, for example, might result in substantially higher damage estimates than had been used when the project was planned.

ECONOMICS 17.

Disregarding income taxes, calculate the rate of return on the following investment opportunities.

A. Invest $100 now.
Receive two payments of $109.46 - one at the end of year 3 and one at the end of year 6.

B. Invest $100 now.
Receive $30.07 at the end of years 1, 2, 3, 4, 5, & 6.

Solution

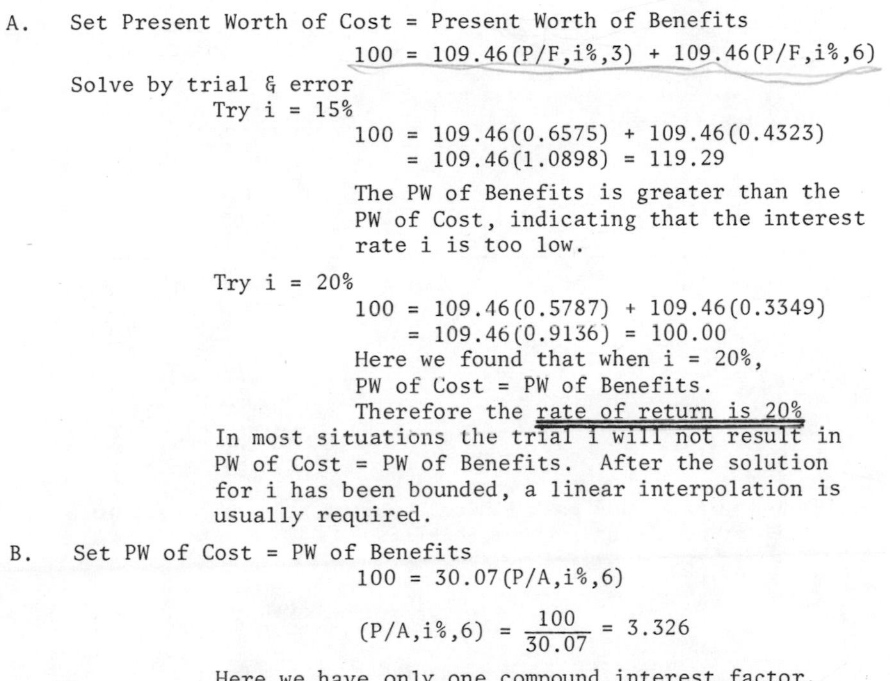

A. Set Present Worth of Cost = Present Worth of Benefits

$$100 = 109.46(P/F,i\%,3) + 109.46(P/F,i\%,6)$$

Solve by trial & error

Try i = 15%

$$100 = 109.46(0.6575) + 109.46(0.4323)$$
$$= 109.46(1.0898) = 119.29$$

The PW of Benefits is greater than the PW of Cost, indicating that the interest rate i is too low.

Try i = 20%

$$100 = 109.46(0.5787) + 109.46(0.3349)$$
$$= 109.46(0.9136) = 100.00$$

Here we found that when i = 20%, PW of Cost = PW of Benefits.

Therefore the rate of return is 20%

In most situations the trial i will not result in PW of Cost = PW of Benefits. After the solution for i has been bounded, a linear interpolation is usually required.

B. Set PW of Cost = PW of Benefits

$$100 = 30.07(P/A,i\%,6)$$

$$(P/A,i\%,6) = \frac{100}{30.07} = 3.326$$

Here we have only one compound interest factor. It can be looked up in interest tables to determine which value of i produces

$$(P/A,i\%,6) = 3.326$$

From interest tables we see that i is exactly 20%

ECONOMICS 18.

A firm expects to receive $32,000 each year for 15 years from the sale of a product. They will have an initial investment of $150,000. Manufacturing and sales expenses will run $7530 per year.

Assume (a) No salvage value and a straight-line write-off of the initial investment over the 15-year period, and

(b) Income tax rate and capital structure are such that income taxes are effectively equal to 48% of taxable income.

Determine the expected rate of return after income taxes.

Solution

This is an excellent example of a simple after-tax rate of return situation. The solution to this type of problem should follow the format indicated on the bottom of Page G-12.

$$\text{Straight line depreciation} = \frac{P - F}{n} = \frac{150{,}000 - 0}{15} = \$10{,}000$$

Year	Before Tax Cash Flow	Straight Line Depreciation	Taxable Income	48% Income Taxes	After Tax Cash Flow
0	-150,000				-150,000
1	+24,470	10,000	14,470	6,946	+17,524
2	+24,470	10,000	14,470	6,946	+17,524
.	.	.	.	.	.
.	.	.	.	.	.
.	.	.	.	.	.
15	+24,470	10,000	14,470	6,946	+17,524

Take the After Tax Cash Flow and compute the rate of return at which PW of Cost equals PW of Benefits.

$$150{,}000 = 17{,}524(P/A,i\%,15)$$

$$(P/A,i\%,15) = \frac{150{,}000}{17{,}524} = 8.559$$

From interest tables, i = 8%

ECONOMICS 19.

Machine A has been completely overhauled for $9000 and is expected to last another 11 years. The $9000 was treated as an expenditure last year. It can be sold now for $20,000 net after selling expenses, but will have no salvage value 11 years hence. It was bought new 9 years ago for $54,000 and has been depreciated by straight line depreciation using a 12 year depreciable life.

Because capacity requirements have diminished, Machine A can be replaced with smaller Machine B. Machine B costs $42,000, has an anticipated life of 20 years, and would reduce operating costs $2500 per year. It would be depreciated the same as Machine A, that is, by straight line depreciation with a 12 year depreciable life.

For an income tax rate of 40% and a capital gains tax of 25%, compare the after-tax annual cost of the two machines and choose the more economical. Use a 10% after-tax rate of return in the calculations.

Solution

Book value of Machine A now = Cost - Depreciation taken to date
= \$54,000 - 9(54,000 - 0)/12 = \$13,500

Long Term Capital Gain on disposal if sold now
= \$20,000 - 13,500 = \$6,500

Tax on capital gain = 0.25(6,500) = \$1,625

Machine A depreciation

$$\frac{P - F}{n} = \frac{54,000 - 0}{12} = 4,500 \text{ per year}$$

Alternate 1 - Keep Machine A for 11 more years

Year	Before Tax Cash Flow	Straight Line Depreciation	Taxable Income	40% Income Taxes	After Tax Cash Flow
0	-20,000		+6,500*	+1,625*	-18,375
1	0	4,500	-4,500	-1,800	+1,800
2	0	4,500	-4,500	-1,800	+1,800
3	0	4,500	-4,500	-1,800	+1,800
4-11	0	0	0	0	0

*Capital gain

After tax annual cost = [18,375 - 1800(P/A,10%,3)][(A/P,10%,11)]
= [18,375 - 1800(2.487)][0.1540]
= \$2,140

Alternate 2 - Buy Machine B

Machine B depreciation

$$\frac{P - F}{n} = \frac{42,000 - 0}{12} = \$3,500 \text{ per year}$$

Year	Before Tax Cash Flow	Straight Line Depreciation	Taxable Income	40% Income Taxes	After Tax Cash Flow
0	-42,000				-42,000
1	+2,500	3,500	-1,000	-400	+2,900
2-12	+2,500	3,500	-1,000	-400	+2,900
13-20	+2,500	0	+2,500	+1,000	+1,500

The annual cost equation may be written in several different forms. One way is:

After tax annual cost = [42,000-1400(P/A,10%,12)][A/P,10%,20] -1500
= [42,000-1400(6.814)][0.1175] -1500
= \$2,314

One aspect of the problem solution requires comment. In Alternate 1 the cash flow in Year 0 reflects the loss of income after capital gains tax from not selling Machine A. This is the preferred way to handle the current market value of the "defender" Machine A.

Choose the Alternative with the smaller annual cost - Keep A

ECONOMICS 20.

Given the following data for two machines:

	A	B
Original Cost	\$55,000	\$75,000
Annual Operation Expense	9,500	7,200
Annual Maintenance Expense	5,000	3,000
Property Taxes, Insurance, and Other Annual Costs	1,700	2,250
Total Annual Costs	16,200	12,450

REQUIRED:

(a) With an interest rate of 10%, at what service life do these two pieces of equipment have an equivalent uniform annual cost?

(b) If the interest rate was 6%, what effect would this have on the average service life at which these machines were equivalent?

(c) Give reasons or show computations to support your answer in (b).

Can also used annual cost.

Solution

(a) The problem may be most readily solved by setting the Present Worth of Cost of Machine A equal to the Present Worth of Cost of Machine B and solving for the unknown service life, N.

$$PW_A = PW_B$$

$$55{,}000 + 16{,}200(P/A,10\%,N) = 75{,}000 + 12{,}450(P/A,10\%,N)$$

$$(P/A,10\%,N) = \frac{75{,}000 - 55{,}000}{16{,}200 - 12{,}450} = \frac{20{,}000}{3{,}750} = 5.33$$

From the 10% compound interest table:

$$(P/A,10\%,8yrs) = 5.335$$

Therefore, the equivalent uniform annual costs are the same at a service life of 8 years.

(b) At a 0% interest rate the breakeven service life would be $\frac{\$20{,}000}{\$3{,}750} = 5.33$ years.

(At 0% interest the Series Present Worth factor equals N.) Thus Machine A is preferred if the actual service life is expected to be less than 5.33 years.

At a 10% interest rate the effect of future annual costs is reduced, with the result that Machine A is preferred for an expected service life not in excess of 8 years. (For an expected service life of over 8 years, of course, Machine B is preferred.)

(c) Following the logic of (a), we see that the problem has been changed so that now (P/A,6%,N) = 5.33

From compound interest tables: (P/A,6%,6yrs) = 4.92
(P/A,6%,7yrs) = 5.58

By linear interpolation the service life at which the machines have equivalent annual costs is:

$$6 + \frac{5.33 - 4.92}{5.58 - 4.92} = 6.6 \text{ years}$$

ECONOMICS 21.

You have to analyze the economics of four plans available to you for the expansion of your firm's manufacturing facilities. Your market studies indicate that maximum sales under all plans are reasonably assured. Revenues and expenses shown in the tabulation below can be used for analysis with a very high probability of their reflecting the true market and investment potential.

Minimum after-income-tax return that you can accept on your 100% equity investment is 6%. Life of the equipment is estimated at 20 years. This life is also to be used for computing income tax depreciation on a straight line basis. Income taxes are to be computed at an assumed 50% rate. The equipment will have no value at the end of its life.

	Plan I	Plan II	Plan III	Plan IV
Investment: Original cost of equipment	$200,000	$250,000	$350,000	$500,000
Annual Gross Revenue	44,800	60,200	101,300	117,900
Total Operating Expenses, excluding income taxes	20,000	27,500	50,200	52,500
Net Annual Revenue:	$24,800	32,700	51,100	65,400

REQUIRED: Analyze the four plans, giving consideration to both total and incremental investment. Recommend which one, if any, should be adopted. Are there other factors that should be considered?

Compute the After Tax Rate of Return for each Plan

Where income taxes are to be considered, the normal procedure is as follows:

1. Compute the Net Annual Revenue before income taxes.
2. Compute the schedule of depreciation charges for tax purposes. In this problem straight line depreciation has been specified.
3. Taxable income is the net annual revenue minus the depreciation.
4. Income Tax is the tax rate (50%) times the taxable income.
5. After Tax Cash Flow is the net annual revenue minus income taxes.
6. Knowing the Investment and the After Tax Cash Flow, we can set:

   Present Worth of Cost = Present Worth of Benefits

   In this problem, with a constant After Tax Cash Flow,

   Investment = After Tax Cash Flow x (P/A,i%,20yrs)

   This equation is solved for i, the after tax rate of return.

These steps for a single alternative are normally organized in horizontal tabular form. This multiple alternative problem may be more conveniently presented as a vertical table.

Investment	Plan I 200,000	Plan II 250,000	Plan III 350,000	Plan IV 500,000	
Net Annual Revenue	$24,800	$32,700	$51,100	$65,400	(1)
Annual Depreciation $= \frac{\text{Investment}}{\text{20 yrs}}$	10,000	12,500	17,500	25,000	(2)
Taxable Income = (1) - (2)	14,800	20,200	33,600	40,400	(3)
Income Taxes = 50% of (3)	7,400	10,100	16,800	20,200	(4)
After Tax Cash Flow = (1) - (4)	17,400	22,600	34,300	45,200	(5)
Ratio: $\frac{\text{Investment}}{\text{After Tax Cash Flow}}$	11.49	11.06	10.20	11.06	

Since the ratio $\frac{\text{Investment}}{\text{After Tax Cash Flow}}$ equals (P/A,i%,20yrs) in this problem, interpolate in the compound interest tables to obtain i - the after tax rate of return.

	Plan I	Plan II	Plan III	Plan IV
(P/A,i%,20yrs) =	11.49	11.06	10.20	11.06
After Tax Rate of Return	$<$ 6%	6.5%	7.5%	6.5%

At this point we have computed the overall rate of return for each alternative. We must compare the overall rate of return for each alternative with the minimum acceptable rate of return (6% in this problem) and reject any alternatives that fail to meet this criterion. Since Plan I yields less than 6%, it should be rejected.

Thus we are left with 3 mutually exclusive alternatives. Which one of the three should be selected? At this point in the analysis we do not know. It would be incorrect to conclude that the Plan with the greatest rate of return is the alternative that should be selected.

The proper analysis method is to examine each separable increment of investment (incremental analysis) to see if it provides a satisfactory increment of benefits. Or stated another way: Each separable increment of investment must produce the minimum acceptable rate of return. If it does not yield this rate of return the increment of investment should not be made.

Incremental analysis is as follows:

A. Rank order the remaining alternatives (after discarding any that fail to meet the overall rate of return criterion) in order of increasing investment.

B. Beginning with the lowest acceptable investment (Plan II), compute the incremental investment and incremental benefits between it (Plan II) and the next higher investment (Plan III).

	Plan II	Plan III
Investment	$250,000	$350,000
After Tax Cash Flow	22,600	34,300

Incremental Investment = \$350,000 - 250,000 = \$100,000
Incremental Benefits = 34,300 - 22,600 = 11,700

$$\text{Ratio } \frac{\Delta \text{ Investment}}{\Delta \text{ After Tax Cash Flow}} = \frac{100,000}{11,700} = 8.55 = (P/A,i\%,20yrs)$$

By interpolation, Incremental rate of return = 9.9%

Since the incremental rate of return exceeds the minimum acceptable rate of return (6%), the incremental investment in an attractive one. Thus if the choice were limited to either Plan II or Plan III, we would select Plan III.

C. Take the outcome from Step B. In this analysis the increment of investment was justified, but in other analyses this may or may not be the case.

Go to the next higher investment alternative (Plan IV) and compare it to the outcome from Step B.

	Plan III	Plan IV
Investment	$350,000	$500,000
After Tax Cash Flow	34,300	45,200

Incremental Investment = $500,000 - 350,000 = $150,000
Incremental Benefits = 45,200 - 34,300 = 10,900

$$\text{Ratio } \frac{\Delta \text{Investment}}{\Delta \text{After Tax Cash Flow}} = \frac{150{,}000}{10{,}900} = 13.76 = (P/A,i\%,20\text{yrs})$$

By interpolation, Incremental rate of return is 3.9%

Here the incremental rate of return is less than the minimum acceptable rate of return, hence the increment of investment is not acceptable. In choosing between Plan III and Plan IV we would select Plan III.

D. The incremental analysis would be continued, always comparing the best one of the alternatives examined to date with the next larger investment alternative. In this problem there is nothing beyond Plan IV, so the analysis has been completed.

Conclusion: Select Plan III.

Other factors to be considered:

Possible constraints that may not have been considered.
- The ability to finance the required investment

Long term marketing considerations.
- Each alternative results in a different share of the total market for this firm.

Intangible considerations
- Safety, etc.

10

OTHER PROBLEMS

PROBLEM 1.

The schematic diagram below shows the operating principle of a transducer used for taking voltage measurements in an alternating current power system. The direct current component of the output is separated from the ripple components by means of a low pass filter and serves as the input for such devices as recorders, indicating instruments, controllers, and telemetering systems.

The transducer is calibrated for operation in terms of the magnitude of the effective (root mean square) value of a sine wave input. The secondary voltage, v_2, and the resistance of R are both made large so that the diode voltage drop causes a negligible error in linearity over the range used.

REQUIRED:

Determine the maximum possible error in measuring the effective value of a wave containing only the fundamental component and a third harmonic component having an amplitude equal to one-tenth that of the fundamental. The phase of the third harmonic component with respect to the fundamental shall be selected to yield the largest error.

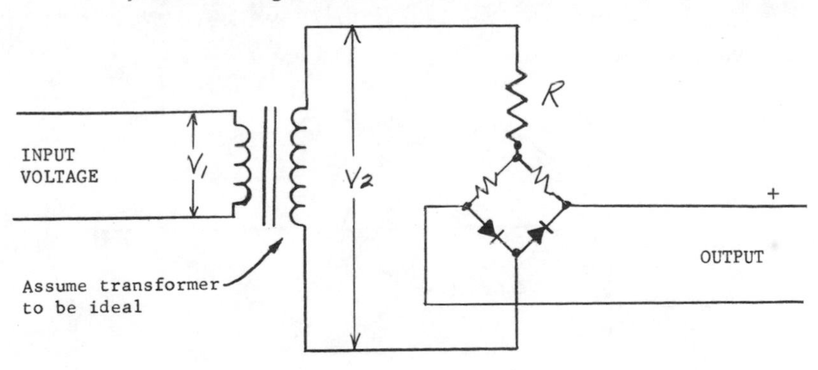

For a pure sine wave input, the output will be as follows:

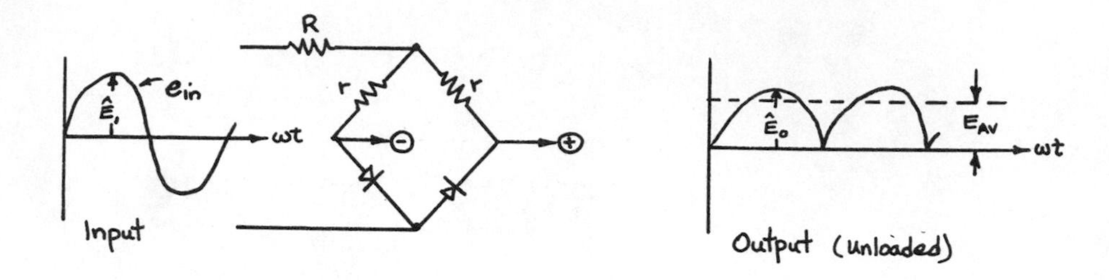

where $\hat{E}_o = \frac{r}{R+r}\hat{E}_i$ (neglecting the diode loss)

$$E_{AV} = \frac{1}{\pi}\int_0^{\pi} \hat{E}_o \sin \omega t \, d\omega t = \frac{2}{\pi}\hat{E}_o$$

Now if the input contains 10% 3rd harmonic:

$$e_{in} = \hat{E}_1 \sin \omega t + (0.1\,\hat{E}_1) \sin 3(\omega t + \phi)$$

then the d.c. average of the output will be dependent upon the change of area of the resulting rectified wave. This area change will depend on the phase of the 3rd harmonic. From consideration of the following diagram it is obvious* that the

---

* Note that if ϕ were 90° the net contribution (in area) due to the 3rd harmonic is zero:

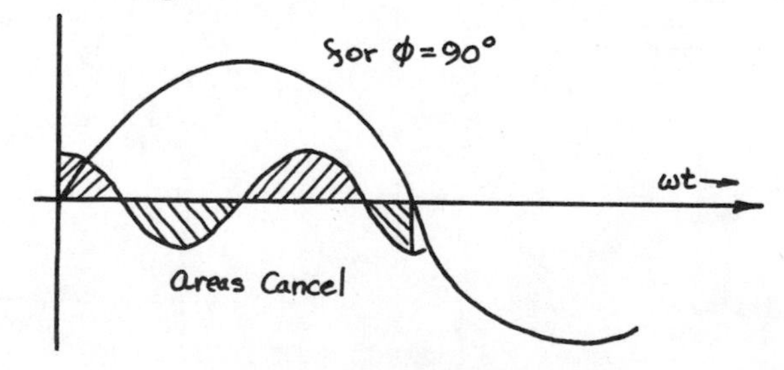

greatest contribution of net area change is when $\phi = 0°$ (or 180°):

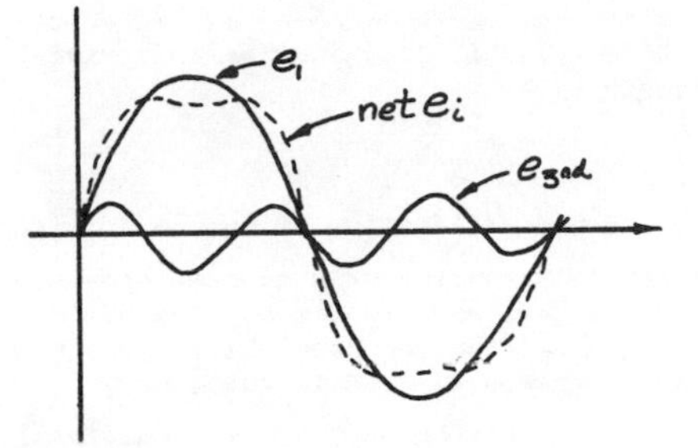

Note that the total area (per half cycle) is equal to the area of the fundamental plus an area of a half cycle of the 3rd harmonic (the next half and next following half cancel each other).

Therefore the new d.c. average will be:

$$E'_{AV} = E_{AV_{fund.}} + \frac{1}{3} E_{AV_{3rd}}$$

$$= \frac{2}{\pi} \hat{E}_{o_{fund}} + \frac{1}{3}\left(\frac{2}{\pi}\right) \hat{E}_{o_{3rd}}$$

$$= \frac{2}{\pi} \hat{E}_{o_{fund}} + \frac{2}{3\pi}\left(0.1\, \hat{E}_{o_{fund}}\right)$$

$$= \left(1 + \frac{0.1}{3}\right) \frac{2}{\pi} \hat{E}_{o_{fund}}$$

Percent change of dc reading is proportional to change in average value:

$$\%\ \text{Change} = 100 \frac{\Delta}{E_{AV}} = \frac{E'_{AV} - E_{AV}}{E_{AV}}(100)$$

$$= \frac{\left(1 + \frac{0.1}{3}\right) \frac{2}{\pi} \hat{E}_o - \frac{2}{\pi} \hat{E}_o}{\frac{2}{\pi} \hat{E}_o}(100)$$

$$= \frac{0.1}{3}(100) = \frac{10}{3} = 3.33\%$$

PROBLEM 2.

A basic electrical system is composed of two subsystem elements, A and B in series, with reliabilities over an operational period of 0.95 and 0.80 respectively.

REQUIRED:

(a) What is the reliability of the basic system?

(b) If one additional A or B element only can be added to the basic system, what is the maximum overall system reliability that can be achieved, and which element would you choose? Show diagrams and calculations.

(c) If any combination of only two additional elements can be added to the basic system, what is the new maximum overall system reliability that can be achieved, and which elements would you choose? Show diagrams and calculations.

(a) System reliability = product of subsystem reliabilities
= (0.95)(0.80) = 0.76

(b) Two alternatives are possible:

Probability of a failure of both B subsystems = (0.20)(0.20) = 0.04

Reliability of dual B subsystems
= 1.00 - 0.04 = 0.96

System reliability = (0.95)(0.96)
= 0.912 = 0.91

Reliability of dual A subsystems
= 1.00 - (0.05)(0.05) = 0.997

System reliability = (0.997)(0.80)
= 0.798 = 0.80

As one would expect, the spare subsystem should backup the less reliable subsystem B.

(c) There are three possible combinations:

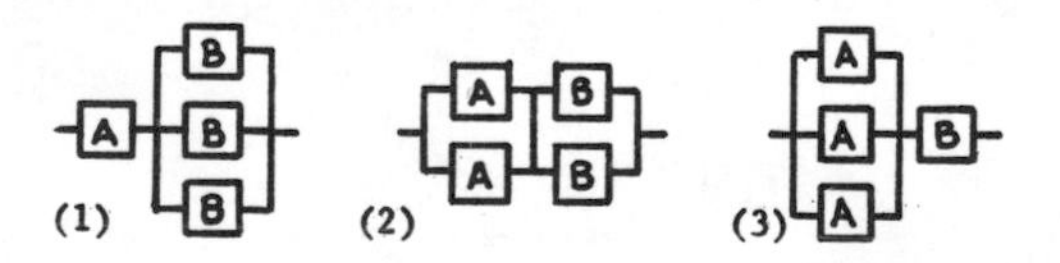

Reliability of B with two backup B subsystems
$= 1.00 - (0.20)(0.20)(0.20) = 0.992$

Reliability of A with two backup A subsystems
$= 1.00 - (0.05)(0.05)(0.05) = 0.999$

Combination			
1	System reliability	= (0.95)(0.992)	= 0.94
2	" "	= (0.997)(0.96)	= 0.96
3	" "	= (0.999)(0.80)	= 0.80

Combination 2, with reliability of 0.96, is the best arrangement of subsystems.

PROBLEM 3.

A series of remote stations is being planned to feed data to a large computer. The data are to be sent by the remote stations and recorded on a tape recording unit and then fed to the computer as necessary.

The data are expected to arrive from the remote stations Poisson distributed at an average rate of 10 transmissions from remote stations per hour. The recording time of the data varies exponentially, with a mean time of four minutes.

REQUIRED:

(a) What is the average waiting time for a remote station before the data will begin to record?

(b) A second tape unit including automatic switching equipment is available at a cost of $2.50 per hour. The telephone lines cost 4¢ a minute per line when used. Is the second unit economically warranted? Show sufficient calculations to justify your answer.

(a) This is the single stage, single server queueing model with Poisson arrivals and exponential service. Any basic operations research text gives the desired queue equations (e.g. Sasieni, Yaspan, and Friedman: Operations Research - Methods and Problems, John Wiley, p126-138).

mean arrival rate $\lambda = 10$ transmissions/hour

mean service time $= 1/\mu = 4$ minutes $= 1/15$ hour

mean service rate $\mu = 15$ recordings/hour

Average waiting time of an arrival

$$E(w) = \frac{\lambda}{\mu(\mu-\lambda)} = \frac{10}{15(15-10)} = \frac{10}{75} \text{ hour} = 8 \text{ minutes}$$

(b) This is the single stage, two server queueing situation. The various expectations for this model may also be obtained from a basic operations research (or queueing) text.

Before proceeding, a quick check can be made. We know a second recorder will substantially reduce (but not eliminate) waiting time. If the elimination of waiting time would not justify the second recorder then we need not bother to make the exact computation. Instead, we could simply conclude the second recorder is not economically warranted.

hourly saving (assuming elimination of waiting time)
= mean arrival rate x mean waiting time reduction x line charge

$$= \lambda\left[E(w_1) - E(w_2)\right] \times 0.04 = (10)(8-0)(0.04) = \$3.20$$

hourly cost = $2.50

Thus we have been unable to show that the second recorder is uneconomical at zero waiting time. We must, therefore, proceed to compute the expectation of average waiting time.

$$P_o = \frac{1}{\left[\sum_{n=0}^{k-1} \frac{1}{n!}\left(\frac{\lambda}{\mu}\right)^n\right] + \frac{1}{K!}\left(\frac{\lambda}{\mu}\right)^K \frac{k\mu}{k\mu-\lambda}}$$

For two service facilities:

$k = 2 \qquad \lambda = 10 \qquad \mu = 15$

$$P_o = \frac{1}{\frac{\lambda}{\mu} + \frac{1}{2}\left(\frac{\lambda}{\mu}\right)^2 \frac{2\mu}{2\mu-\lambda}} = \frac{1}{\frac{10}{15} + \frac{100}{450}\cdot\frac{30}{20}} = \frac{1}{1} = 1$$

Average waiting time of an arrival

$$E(w) = \frac{\mu\left(\frac{\lambda}{\mu}\right)^k}{(k-1)!(k\mu-\lambda)^2} P_o = \frac{15\cdot\left(\frac{10}{15}\right)^2}{(30-10)^2}(1) = \frac{1}{60} \text{ hour}$$

$$= 1 \text{ minute}$$

hourly saving of second recorder $= \lambda\left[E(w_1) - E(w_2)\right] \times 0.04$

$$= (10)(8.0 - 1.0)(0.04) = \$2.80$$

Thus the second recorder is economically warranted, for the hourly saving exceeds the hourly cost.

PROBLEM 4.

Switch Reliability

As a manufacturer of high quality equipment you need a toggle switch which must have a reliability of 1,000,000 operations between failures. Switch failure in this case means failure to operate in an acceptable manner, rather than catastrophic failure.

A potential supplier gives you the following manufacturers' test data.

Switch Number	Millions of Test Operations	Number of Failures
1	2.0	2
2	1.5	1
3	3.0	0
4	1.0	2
5	2.5	0
6	2.0	0
7	1.5	2
8	1.5	1
9	2.0	0
10	1.0	3

REQUIRED:

Are the switches acceptable? State your confidence level and your reason for choosing it.

n (millions)	x
2.0	2
1.5	1
3.0	0
1.0	2
2.5	0
2.0	0
1.5	2
1.5	1
2.0	0
1.0	3
$\Sigma n = 18.0$	$\Sigma x = 11$

Since the reliability given is one that should not be exceeded, a one-sided confidence limit is needed. It does not require an extremely high level of confidence, since the consequences of failure are given as non-catastrophic. A 90% conficence level could be used. However, the normal approximation to the binomial is less accurate for values of p near 0 or 1, so a more stringent 95% level will be used to compensate for the inaccuracy of the approximation.

$$\bar{p} = \frac{\Sigma x}{\Sigma n} = \frac{11}{18 \text{ (millions)}}$$

$$\text{Upper Confidence Limit} = \bar{p} + Z_{.05}\sqrt{\frac{\bar{p}(1 - \bar{p})}{N}}$$

where $N = \Sigma n = 18 \times 10^6$

$$\bar{p} = \frac{11}{18} \times 10^{-6}$$

$$1 - \bar{p} \simeq 1$$

$$Z_{.05} = 1.645$$

$$\text{U. C. L.} \simeq 0.611 \times 10^{-6} + 1.645\sqrt{\frac{11 \times 1}{18 \times 10^6 \times 18 \times 10^6}}$$

$$\simeq 0.611 \times 10^{-6} + 0.303 \times 10^{-6}$$

$$\simeq 0.914 \times 10^{-6} < 1 \text{ per } 1{,}000{,}000 \text{ operations.}$$

With about 95% confidence, the reliability is within the required value. Therefore, the switches are acceptable.

PROBLEM 5.

Telephone Statistical Analysis

A study has been made of the operation of a portion of a telephone system, and the following samples of data have been collected:

Sample circuit holding times of 10, 50, 30, 250, 190, 50, 70, 130, 40, and 280 seconds.

326 sample operator work times for a group of PBX operators to handle a type of call classified as "extension to central office trunk," for which the sample mean was 29.7 seconds, with a standard deviation of 14.2 seconds.

REQUIRED:

(a) Compute the mean and the standard deviation of the sample of the circuit holding times indicated above.

(b) With 90% confidence, calculate the limits of the interval (confidence interval) which contains the actual mean of the PBX operator's work times for the class of call indicated above.

X_i	X_i^2
10	100
50	2500
30	900
250	62500
190	36100
50	2500
70	4900
130	16900
40	1600
280	78400
1100	206,400

$$s^2 = \frac{n\sum X_i^2 - (\sum X_i)^2}{n(n-1)}$$

$$= \frac{10(206{,}400) - (1100)^2}{10(9)}$$

$$= \frac{2{,}064{,}000 - 1{,}210{,}000}{90} = \frac{854{,}000}{90}$$

$$s = \sqrt{\frac{85{,}400}{9}} = \frac{29.22}{3} = 9.74$$

$$\overline{X} = \frac{\sum X_i}{n} = \frac{1100}{10} = 110$$

(a) Mean = 110
Standard Deviation = 9.74

(b) $n = 326$ $\overline{X} = 29.7$ $s = 14.2$

$$C.\ I. = (\overline{X} - Z_{.05}\, s/\sqrt{n},\ \overline{X} + Z_{.05}\, s/\sqrt{n})$$

$$= (29.7 - (1.645)(14.2)/\sqrt{326},\ 29.7 + (1.645)(14.2)/\sqrt{326})$$

$$= (29.7 - 1.3,\ 29.7 + 1.3)$$

$$= (28.4,\ 31.0)$$

PROBLEM 6.

Computer Programming

If the lengths of the sides of a triangle are given by the values of the variables x, y, and z, then the area of the triangle can be computed from:

$$\text{Area} = \sqrt{w(w-x)(w-y)(w-z)} \qquad \text{where} \quad w = \frac{x + y + z}{2}$$

You are to read the values x, y, and z from a standard keypunched card.

No value of x, y, or z exceeds 99.

The value of x is punched in columns 1 to 10, with a decimal point, but without an exponent.

The value of y in columns 11 to 20 and the value of z in columns 21 to 30 are in the same form as the value of x.

After the Area has been computed, the Area and the lengths of the sides are to be printed, with the Area printed last. The printer has a capacity of 128 characters per line. Each of the four numbers will occupy 16 printing positions and will have eight decimal places and an exponent.

REQUIRED:

Write a program that will perform the above computation using FORTRAN language. Number on your page for 1 to 72 positions; 1 to 5 for statement number, 6 blank, and 7 to 72 for the statements themselves.

A FORTRAN program might be as follows:

```
STATEMENT
 NUMBER   FORTRAN STATEMENT
      READ(60,10) X,Y,Z
   10 FORMAT(3F10.7)
      W=(X+Y+Z)/2.
      AREA=SQRT(W*(W-X)*(W-Y)*(W-Z))
      WRITE(61,20)
   20 FORMAT(6X6HSIDE X,14X6HSIDE Y,14X6HSIDE Z,15X4HAREA
      WRITE(61,30) X,Y,Z, AREA
   30 FORMAT(1X,4(E16.8,4X))
      END
```

Below is a data card to test the program.

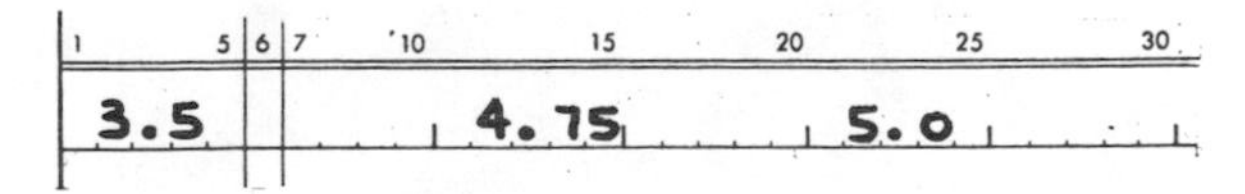

The keypunched program and data card are as follows:

```
      READ(60,10) X,Y,Z
   10 FORMAT(3F10.7)
      W=(X+Y+Z)/2.
      AREA=SQRT(W*(W-X)*(W-Y)*(W-Z))
      WRITE(61,20)
   20 FORMAT(6X6HSIDE X,14X6HSIDE Y,14X6HSIDE Z,15X4HAREA)
      WRITE(61,30) X,Y,Z, AREA
   30 FORMAT(1X,4(E16.8,4X))
      END

 3.5       4.75      5.0
```

Below is the output from the program.

```
      SIDE X              SIDE Y              SIDE Z               AREA
  3.50000000E 00      4.75000000E 00      5.00000000E 00      7.94228141E 00
```

PROBLEM 7.

Subject: Computer: Logic and Circuitry

Two binary coded digits, A and B, are set into the toggles (flip-flops) shown symbolically in the accompanying diagram. Each digit is composed of two bits, "1"s and "2"s, and can, therefore, have any value from zero through three.

The outputs at the top of each toggle have two possible voltage levels, -6.0 volts and approximately 0.0 volts. Each toggle set to the "1" state (or ON) has the "1" side output high (approx. 0.0 volts) and the "0" side output low (approx. -6.0 volts).

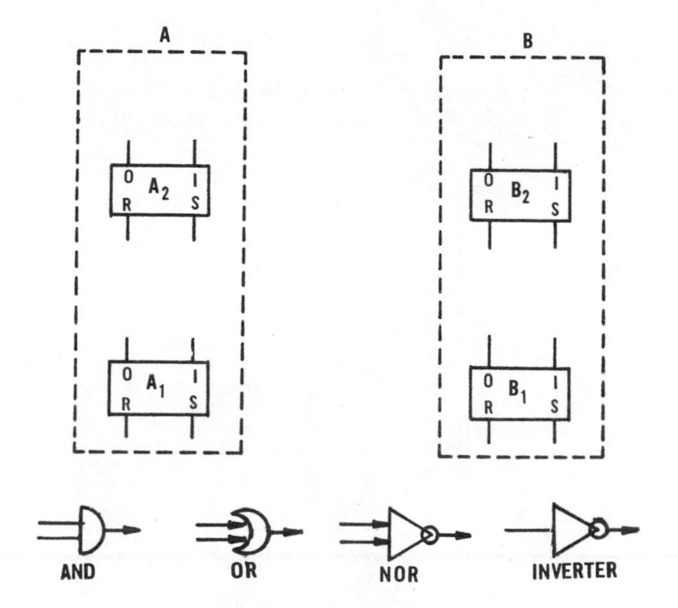

(1) Wt. 6 Using the symbols given, show the logical gating and connections necessary to compare the digits (A and B) and to produce a high level (approx. 0.0 volts) output when, and only when, A is greater than B. Any usable type of logic may be assumed, such as diode logic, resistor-transistor logic, etc., or any combination of methods. Minimize the gates and components used as much as possible. Use the logical symbols shown for "AND", "OR", "NOR" gates or "INVERTERS". Or, you may use other, or additional, symbols if they are labelled and also shown in part (2), below.

(2) Wt. 4 Sketch a schematic diagram typical of each type of gate used in the above logic. Assume a positive and negative supply voltage and ground. Show any gate returns to either supply or ground, but omit all values.

The following truth table expresses the wording of the problem:

	A		B		
	A_2	A_1	B_2	B_1	f
0	0	0	0	0	0
1	0	1	0	1	0
2	0	0	1	0	0
3	0	0	1	1	0
4	0	1	0	0	1
5	0	1	0	1	0
6	0	1	1	0	0
7	0	1	1	1	0
8	1	0	0	0	1
9	1	0	0	1	1
10	1	0	1	0	0
11	1	0	1	1	0
12	1	1	0	0	1
13	1	1	0	1	1
14	1	1	1	0	1
15	1	1	1	1	0

The output "f" is to have a value - must be a "1" - if and only if the digit "A" is larger than the digit "B". The function "f" expressed in Minterm Canonical form would be (from the truth table):

$$f = M_4 + M_8 + M_9 + M_{12} + M_{13} + M_{14}$$

$$= \bar{A}_2A_1\bar{B}_2\bar{B}_1 + A_2\bar{A}_1\bar{B}_2\bar{B}_1 + A_2\bar{A}_1\bar{B}_2B_1 + A_2A_1\bar{B}_2\bar{B}_1$$

$$+ A_2A_1\bar{B}_2B_1 + A_2A_1B_2\bar{B}_1$$

If we choose to work with this type of expression, then the minimal form of the function can be found by a variety of techniques. The laws of Boolean algebra may be applied to obtain some reduction, a Karnaugh map or a Veitch diagram may be used to obtain the minimal form.

However, a better approach, especially for a more difficult problem, would be as follows (the Minterms correspond to the vertices of a N dimensioned cube, the covering of the cube gives the minimal form of the function):

f = T

4 $\quad 4/12 \rightarrow (4,3) \rightarrow A_1B_2B_1$

8

12 $\quad 8/9, 12/13 \rightarrow (8,2) \rightarrow A_2B_2$

$\quad 14/12 \rightarrow (12,1) \rightarrow A_2A_1B_1$

13

14

$$f_{min} = A_1\overline{B}_2\overline{B}_1 + A_2\overline{B}_2 + A_2A_1\overline{B}_1 \qquad (1)$$

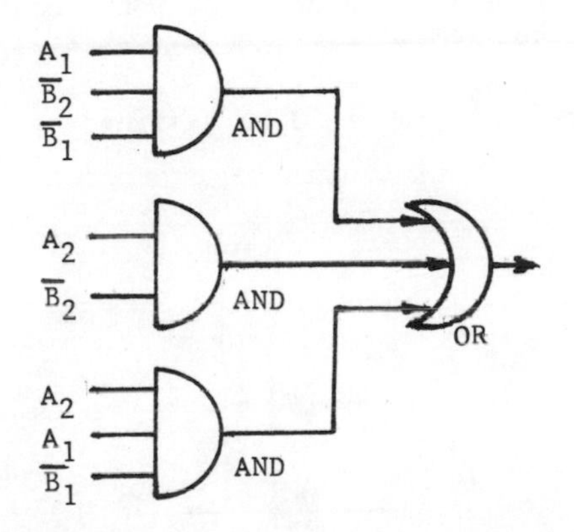

A second answer would be obtained by factoring the function as follows:

$$f = A_1\overline{B}_1 \left[\overline{B}_2 + A_2\right] + A_2\overline{B}_2 \qquad (2)$$

This would add another stage of signal reduction and delay.

(2) Working with expression (1) of the first part of this problem, the AND gates would be designed as follows:

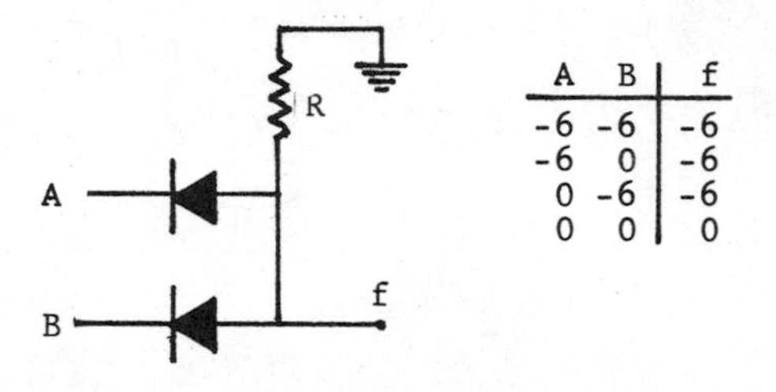

A	B	f
-6	-6	-6
-6	0	-6
0	-6	-6
0	0	0

With the above circuit, there would be an output if, and only if, both input "A" and input "B" were at the 0 voltage level.

The OR gate would be designed as follows:

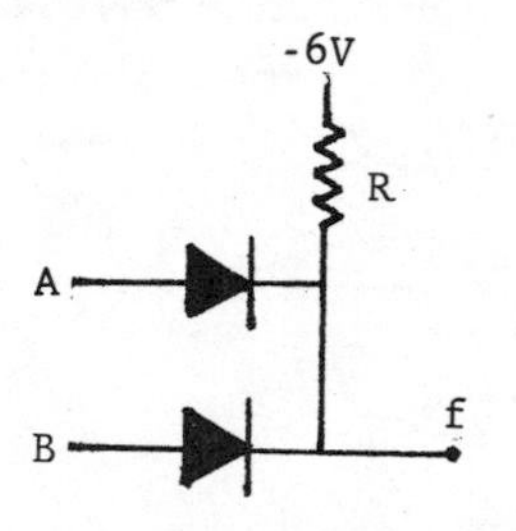

This circuit would have an output if either one or both inputs were at a zero voltage level.

PROBLEM 8.

Resolve the system $E_A = 1500 \angle 30°$, $E_B = 1800 \angle -70°$, $E_C = 2000 \angle 170°$ into its symmetrical components.

$$E_{A1} = \frac{1}{3}(1500 \angle 30° + 1800 \angle -70° + 120° + 2000 \angle 170° + 240°)$$
$$= \frac{1}{3}(1500 \times 0.866 + j1500 \times 0.5 + 1800 \times 0.642 + j1800 \times 0.766$$
$$+ 2000 \times 0.642 + j2000 \times 0.766)$$
$$= \frac{1}{3}(1300 + j750 + 1156 + j1379 + 1284 + j1532)$$
$$= \frac{1}{3}(3740 + j3661) = 1246 + j1220 = 1740 \angle 44°$$
$$E_{B1} = 1740 \angle 44° + 240° = 1740 \angle 284°$$
$$E_{C1} = 1740 \angle 44° + 120° = 1740 \angle 164°$$
$$E_{A2} = \frac{1}{3}(1500 \angle 30° + 1800 \angle -70 + 240° + 2000 \angle 170 + 120°$$
$$= \frac{1}{3}(1500 \times 0.866 + j1500 \times 0.5 - 1800 \times 0.985 + 1800 \times 0.174$$
$$+ 2000 \times 0.342 - j2000 \times 0.94)$$
$$= \frac{1}{3}(1300 + j750 - 1773 + j313 + 684 - j1880)$$
$$= \frac{1}{3}(211 - j817) = 70 - j272 = 281 \angle 284° = 281 \angle -76°$$
$$E_{B2} = 281 \angle -76° + 120° = 281 \angle 44°$$
$$E_{C2} = 281 \angle -76° + 240° = 281 \angle 164°$$
$$E_{A0} = \frac{1}{3}(1500 \angle 30° + 1800 \angle -70° + 2000 \angle 170°)$$
$$= \frac{1}{3}(1500 \times 0.866 + j1500 \times 0.5 + 1800 \times 0.342 - j1800 \times 0.94$$
$$- 2000 \times 0.985 + 2000 \times 0.174$$
$$= \frac{1}{3}(1300 + j750 + 616 - j1692 - 1970 + j348)$$
$$= \frac{1}{3}(-54 - j594) = 18 - j198 = 199 \angle 266°$$
$$E_{A0} = E_{B0} = E_{C0} = 199 \angle 266°$$

Reference: Elements of Power Systems Analysis by William D. Stevenson, Jr., Chapter 13.

PROBLEM 9.

It is required that a loading of 5 KW be maintained in a heating unit. At an initial temperature of 15°C, a voltage of 220 is necessary for this purpose. After the unit has settled down to a steady state, it is found that a voltage of 240 is required to maintain this loading. The temperature coefficient of the conductor is 0.0006 Ω per 1°C.

REQUIRED:

Calculate the final temperature of the heating element in °C.

The power developed at 220 volts:

$$P = \frac{E^2}{R_1} = \frac{220^2}{R_1} = 5000 \text{ watts}$$

$$R_1 = \frac{220^2}{5000} = 9.68 \text{ ohms}$$

The power developed at 240 volts:

$$P = \frac{E^2}{R_2} = \frac{240^2}{5000} = 5000 \text{ watts}$$

$$R_2 = \frac{240^2}{5000} = 11.52 \text{ ohms}$$

Since: $R_{t2} = R_{t1}\left[1 + \alpha(T_2 - T_1)\right]$

Thus: $R_2 = R_1 + R_1\alpha(T_2 - T_1)$

And $\dfrac{R_2 - R_1}{R_1\alpha} = T_2 - T_1$

$$\frac{11.52 - 9.68}{9.68 \times 0.0006} + 15 = T_2 \quad \text{OR} \quad 317 + 15 = T_2$$

Final temperature of the heating element:

$$T_2 = 332°C$$

PROBLEM 10.

It is desired to deposit a thickness of 1 mm of nickel on the curved surface of a worn steel pin 10 inches in diameter. The available current is 50 amperes. The electro-chemical equivalent of nickel is 0.000308 gm/coulomb, and its relative density is 8.9.

REQUIRED:

Calculate the time required.

Converting, we obtain:

length = 10 in = 25.4 cm

diameter = 4 in = 10.2 cm

radius = 2 in = 5.1 cm

thickness = 1 mm = 0.1 cm

Assuming a non-tapered pin shape:

Area of surface $= 2\pi r\,l = 6.28 \times 5.1 \times 25.4 = 814\ cm^2$

Area of the two ends $= 2\,(r^2\pi) = 2\,(5.1^2 \times 3.14) = \underline{163\ cm^2}$

$977\ cm^2$

Total area of pin:

Volume $= A \cdot$ thickness $= 977 \times 0.1 = 97.7\ cm^3$

Weight of added nickel = Vol.x relative density

$= 97.7 \times 8.9 = 870$ grams

Coulombs required $= \dfrac{870\ gr}{0.00038\ gr/coul} = 229 \times 10^4$ coul

$= 229 \times 10^4$ amp sec

Time required $= \dfrac{2{,}290{,}000\ amp\ sec}{50\ amps}$

= 45,800 sec = 763 min = 12 hrs & 43 min   ANS.

PROBLEM 11.

An electric soldering machine is supplied with 240 volts AC, and draws a current of 0.5 amperes. The weight of the copper soldering tip is 130 grams, and 60 per cent of the heat generated is lost in heating other metal parts of the machine and in radiation.

Melting point of solder = 300°C   Molecular weight of copper = 63.5

Ambient temperature = 15°C   Density of copper = 8.9 grams/CM^3

Specific heat of copper = 0.094 cal/gram/°C

(Assume to be independent of temperature)

REQUIRED:

Determine how much time must elapse after switching on the machine before the soldering tip is heated to the melting point of solder and the machine is ready for use.

Δt = temperature difference = 300°C - 15°C = 285°C

Δt x weight (grams) = 285 x 130 = 37,250 gram-degrees

The required heat = Cal/gram/°C x gram-degrees = Calories

= 0.094 x 37,250 = 3,520 Calories

Heat Saved =	40%	= 3,520
Heat Lost =	60%	= 5,270
Total Heat	100%	= 8,790 Calories

Using the conversion factor of watt seconds/calories = 4.184 (See Standard Handbook for Electrical Engineers - Sec. 1-87: Energy Conversion Tables), we obtain

$$8{,}790 = \frac{E.i.t}{4.184} = \frac{240 \times 0.5 \times t}{4.184}$$

then $$t = \frac{4.184 \times 8{,}790}{740 \times 0.5} = 306 \text{ sec.} = 5.1 \text{ min.}$$ ANS.

PROBLEM 12.

The voltages of an unbalanced 3-phase supply are $V_a = (200 + j0)V$, $V_b = (- j200)V$ and $V_c = (- 100 + j200)V$.

Connected in star across this supply are three equal impedances of (20 + j10) ohms. There is no connection between the star point and the supply neutral.

REQUIRED:

Evaluate the symmetrical components of the A phase current and the three line currents.

The voltage components are as follows:

$$V_1 = \frac{1}{3}(V_a + aV_b + a^2Vc)$$
$$= \frac{1}{3}\left[200 + (-0.5 + j0.866)(- j200) + (-0.5 - j0.866)(- 100 + j200)\right]$$
$$= \frac{1}{3}(200 + j100 + 173.2 + 50 - j100 + j86.6 + 173.2)$$
$$= \frac{1}{3}(596.4 + j86.6) = 198.8 + j28.86$$
$$V_2 = \frac{1}{3}(V_a + a^2V_b + aV_c)$$
$$= \frac{1}{3}\left[200 + (- 0.5 - j0.866)(-j200) + (- 0.5 + j0.866)(- 100 + j200)\right]$$
$$= \frac{1}{3}(200 + j100 - 173.2 + 50 - j86.6 - j100 - 173.2)$$
$$= \frac{1}{3}(- 96.4 - j86.6) = - 32.13 - j28.86$$
$$V_o = \frac{1}{3}(V_a + V_b + V_c) = \frac{1}{3}(200 + j0 - j200 - 100 + j200)$$
$$= \frac{1}{3}(100) = 33.33$$

$V_{a1} = 198.8 + j28.86 = (20 + j10)\ I_{a1}$

$V_{a2} = -32.13 - j28.86 = (20 + j10)\ I_{a2}$

$V_{ao} = 33.33 = \infty.\ I_{ao}$, since neutral is not connected, i.e., there is no connection between the star point and the supply neutral.

$$I_{a1} = \frac{198.8 + j28.86}{20 + j10}$$ ANS.

$$I_{a2} = \frac{-32.13 - j28.86}{20 + j10}$$ ANS.

$$I_{ao} = \frac{33.33}{\infty} = 0$$ ANS.

$$I_a = I_{a1} + I_{a2} + I_{ao} = \frac{1}{20 + j10}(198.8 + j28.86 - 32.13 - j28.86)$$

$$= \frac{1}{20 + j10}(166.67) = \frac{20 - j10}{500} \times 166.67 = I_a = \underline{\underline{6.67 - j3.33 \text{ amps in line a}}}$$

ANS.

To obtain the other (b,c) line currents:

$$I_b = a^2 I_{a1} + aI_{a2} + I_{ao}$$

$$= \left[(-0.5 - j0.866)(198.8 + j28.86) + (-0.5 + j0.866)(-32.13 - j28.86)\right]\frac{1}{20 + j10}$$

$$= (-99.40 - j172.16 - j14.43 + 24.99 + 16.65 - j27.82 + j14.43 + 24.99)\ \frac{20 - j10}{500}$$

$$= (-33 - j200)\ \frac{20 - j10}{500} \quad = \underline{\underline{-5.33 - j7.33 \text{ amps in line b}}}$$ ANS.

$$I_c = a\, I_{a1} + a^2\, I_{a2} + I_{ao}$$

$$= \left[(-0.5 + j0.866)(198.8 + j28.86) + (-0.5 - j0.866)(-32.13 - j28.86)\right]\frac{1}{20 + j10}$$

$$= (-99.40 + 172.16 + j14.43 - 24.99 + 16.65 + j27.82 - j14.43 - 24.99)\ \frac{20 - j10}{500}$$

$$= (-133.33 + j200)\ \frac{20 - j10}{500}$$

$$= \underline{\underline{-1.33 + j10.67 \text{ amps in line c}}}$$ ANS.

As a check: $I_a + I_b + I_c = 0$, or

$6.67 - j3.33 - 5.33 - j7.33 - 1.33 + j10.67 = 0$

Reference: Elements of Power Systems Analysis by William D. Stevenson, Jr. Chapter 13.

PROBLEM 13.

A single sensor has been used to detect excessive levels of a contaminant in a chemical process. Its output is "0" for normal conditions and "1" when the impurity level becomes too high. An evaluation program shows that the sensor is subject to occasional false alarms; also, it sometimes fails to operate when the contaminant level is high.

To improve the situation, it is decided to use three sensors and to disregard the indication of any single sensor whose output differs from the other two sensors. It is desired to have a single "correct signal" output which will be "0" for normal conditions and "1" for an excessive contaminant level, and an "alarm" output which will be "0" when all sensor outputs are identical ("0" or "1") and which will latch into a "1" output at any time when the three sensor outputs become not identical.

Also, an "alarm reset" input is to be provided such than a "1" input to it will reset the alarm output to zero after an alarm condition ceases to exist.

REQUIRED: Derive a logic system to meet the above requirements. The system is to be implemented with NOR logic. There are no "fan in" or "fan out" limitations, i.e. any NOR gate can have as many inputs as may be needed and can drive any necessary number of gates from its output. Your solution should include:

(a) A truth table for the "correct signal" and "alarm" outputs.
(b) The logic equations.
(c) A logic circuit diagram showing how the sensors and NOR gates are to be connected.

Solution

Let sensor outputs be A, B, and C; then a truth table may be formed with f_0 being the contaminant signal and f_A being the alarm.

Term	Sensors			Outputs	
m	A	B	C	f_0	f_A
0	0	0	0	0	0
1	0	0	1	0	1
2	0	1	0	0	1
3	0	1	1	1	1
4	1	0	0	0	1
5	1	0	1	1	1
6	1	1	0	1	1
7	1	1	1	1	0

$$\therefore f_0 = \sum m(3,5,6,7)$$

$$\therefore f_A = \sum m(1,2,3,4,5,6)$$

Then simplify by use of Karnaugh Map.

For f_0:

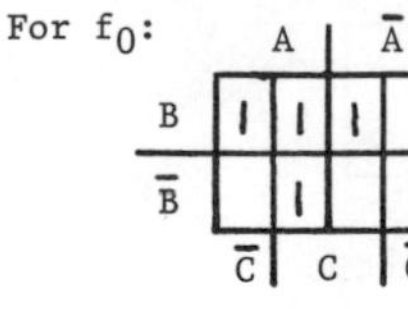

$$\therefore f_0 = AB + BC + AC$$

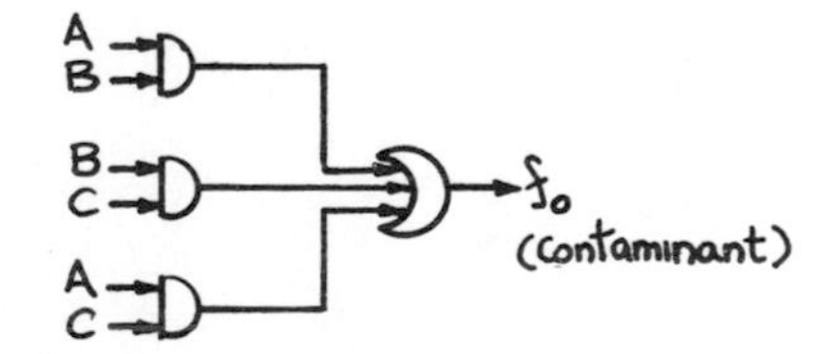

For f_A:

$$\therefore f_A = \bar{A}B + \bar{B}C + A\bar{C}$$

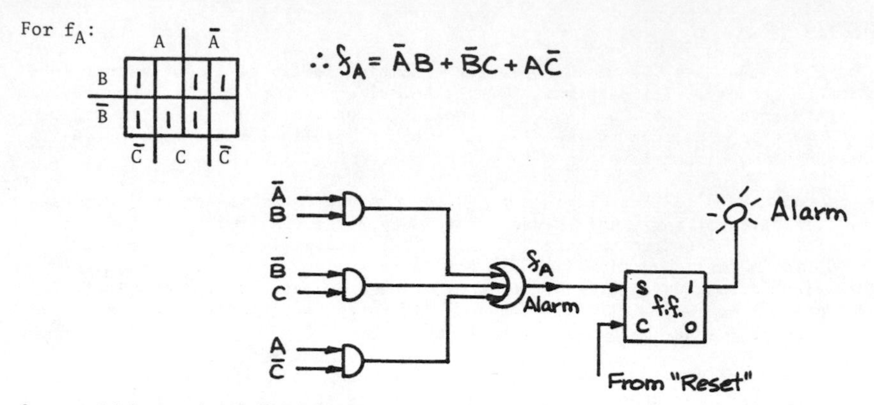

One could have simplified the Karnaugh Map in terms of "O's" as:

$$\bar{f}_o = \bar{B}\bar{C} + \bar{A}\bar{B} + \bar{A}\bar{C}$$

$$\therefore f_o = \overline{\bar{B}\bar{C} + \bar{A}\bar{B} + \bar{A}\bar{C}}$$

$$= (B+C)\cdot(A+C)\cdot(A+B) \text{ to go directly to NOR logic:}$$

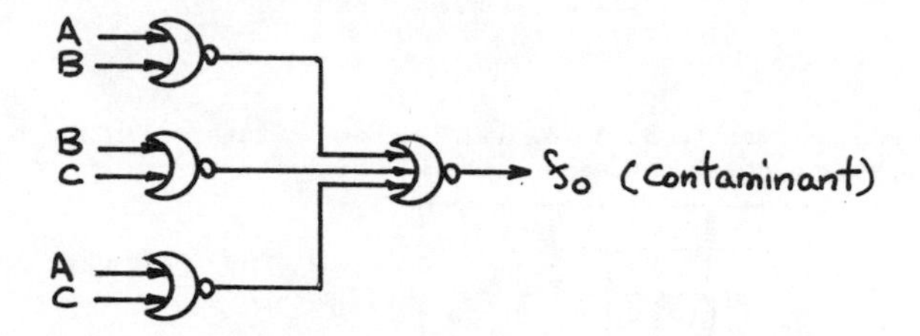

And $\bar{f}_A = ABC + \bar{A}\bar{B}\bar{C}$

$$f_A = \overline{ABC + \bar{A}\bar{B}\bar{C}} = (\bar{A}+\bar{B}+\bar{C})\cdot(A+B+C)$$

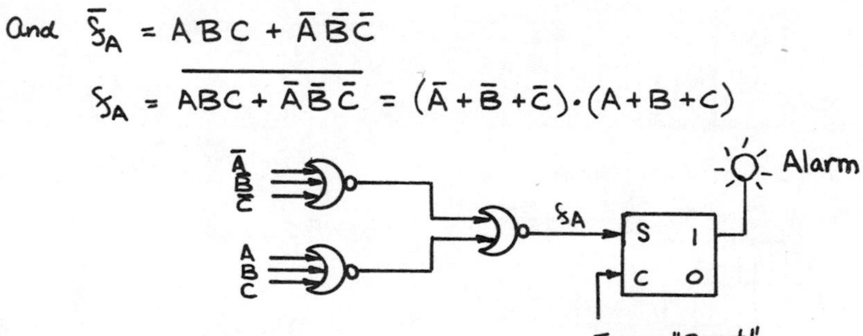

11
TYPICAL EXAMINATION SET

This chapter gives a typical set of problems that could be considered to be fairly equivalent to the NCEE Electrical Engineering examination. The problems are deemed to be the same level as those on the NCEE exam and have been prepared after making a survey of the past exams - including 1974. Like the rest of the problems in this book, these are not actual NCEE Electrical Engineering exam problems.

The examination has been given in two parts, half in the morning and half in the afternoon.

MORNING SESSION

Here are 10 problems representing the morning session. You would be asked to select any 4 problems and to work them in detail showing your method of solution. As in any exam, you might find a poorly stated problem; in any case, you should make any necessary assumptions and carefully state these assumptions.

PROBLEM 1. Circuits. (Usually at least one or more problems involves an RLC transient or an RLC series parallel combination circuit using steady ac circuit analysis.)

The circuit shown on the next page is part of a control system. During a routine maintenance operation, the circuit is accidently broken at point X, and the maintenance person is subject to an electrical shock. Assume that before the break the circuit was operative under steady-state conditions.

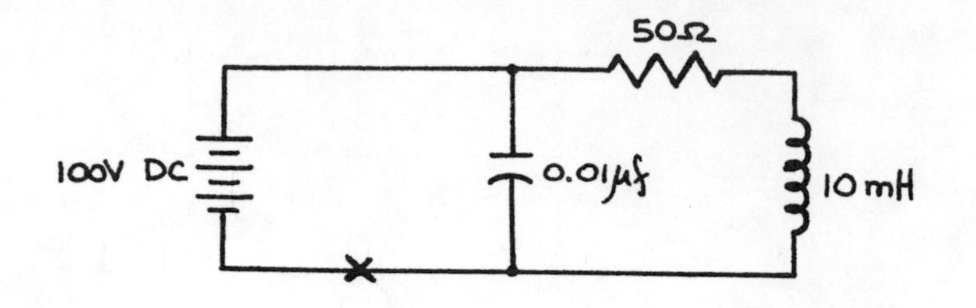

REQUIRED: Derive an expression for the voltage across the break as a function of time. (t = 0 at the instant the break occurred.)

Solution

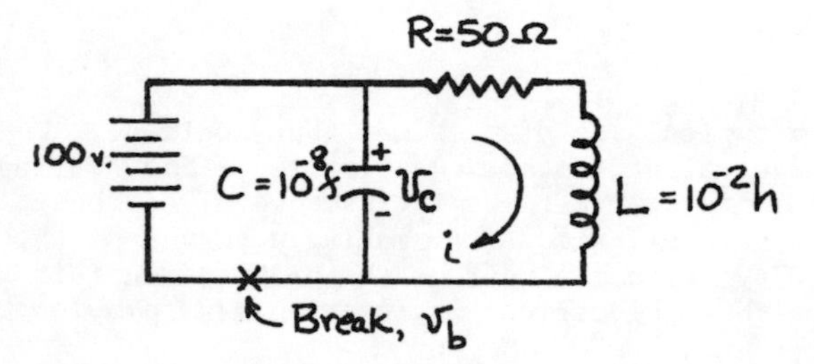

Before "break" the capacitor is fully charged to 100 v and the current $i_L = \frac{V}{R} = \frac{100}{50} = 2\text{ A}$

$$\therefore\ v_c(0) = 100\text{ v} \qquad i_L(0) = 2\text{ A}$$

After "break" the loop equation may be found and then converted to charge, q:

$$v_c + Ri + L\frac{di}{dt} = 0 = q/C + R\dot{q} + L\ddot{q} = 0$$

with $v_c(0) = 100\text{ v}$ or $q(0) = v_c C = 10^{-6}$

$i_L(0) = 2\text{ A}$ or $\dot{q}(0) = 2\text{ A}$

$$\ddot{q} + \frac{R}{L}\dot{q} + \frac{1}{LC}q = 0$$

∴ Roots M_1 and M_2 may be found from quadratic equation:

$$M_{1,2} = -\frac{R}{2L} \pm \frac{1}{2}\sqrt{\left(\frac{R}{L}\right)^2 - \frac{4}{LC}}$$

$$= -\frac{50}{2\times 10^{-2}} \pm \frac{1}{2}\sqrt{\left(\frac{50}{10^{-2}}\right)^2 - \frac{4}{10^{-10}}}$$

$$= -\alpha \pm j\beta \quad \text{where } \alpha = 2.5\times 10^{+3},\ \beta = 10^5$$

$$q = K_1 e^{(-\alpha + j\beta)t} + K_2 e^{(-\alpha - j\beta)t}$$

But $e^{\pm j\beta t} = \cos\beta t \pm j\sin\beta t$

$$\therefore q = e^{-\alpha t}\left[K_3 \cos\beta t + K_4 \sin\beta t\right]$$

where $K_3 = K_1 + K_2$

$K_4 = j(K_1 - K_2)$

@ $t=0$

$$q(0) = 10^{-6} = e^0\left[K_1(1) + K_4(0)\right] = K_3$$

And, for the initial condition, $\dot{q}$ is needed:

$$\dot{q} = (-\alpha)e^{-\alpha t}\left[K_3 \cos\beta t + K_4 \sin\beta t\right]$$
$$+ e^{-\alpha t}\left[(-\beta)K_3 \sin\beta t + (\beta)K_4 \cos\beta t\right]$$

@ $t=0$

$$\dot{q}(0) = 2 = -\alpha e^0\left[K_3 + 0\right] + e^0\left[\beta K_4\right] = -\alpha K_3 + \beta K_4$$

$$= -(2.5\times 10^3)K_3 + 10^5 K_4$$

$$= -2.5\times 10^3(10^{-6}) + 10^5 K_4$$

$$2 + 2.5\times 10^{-3} = 10^5 K_4, \quad K_4 \approx 2\times 10^{-5}$$

$$\therefore q = e^{-2.5\times10^{-3}t}\left[10^{-6}\cos 10^5 t + 2\times10^{-5}\sin 10^5 t\right]$$

And v_b (break voltage) $= 100 - v_c$

where $v_c = q/c = 10^2 e^{-2500t}(\cos 10^5 t + 20 \sin 10^5 t)$

$$\approx 2\times10^3 e^{-2500t}\sin(10^5 t + \phi)$$

where $\phi = \tan^{-1}\frac{1}{20} = 2.86^\circ$

Therefore

$$v_b = 100 - 2\times10^3 e^{-2.5\times10^3 t}\sin(10^5 t + 2.86^\circ)$$

PROBLEM 2. Control Systems. (Many of the NCEE questions ask something about system stability for a linear control system.)

The following system is to be stabilized by the addition of tachometer feedback, K_t.

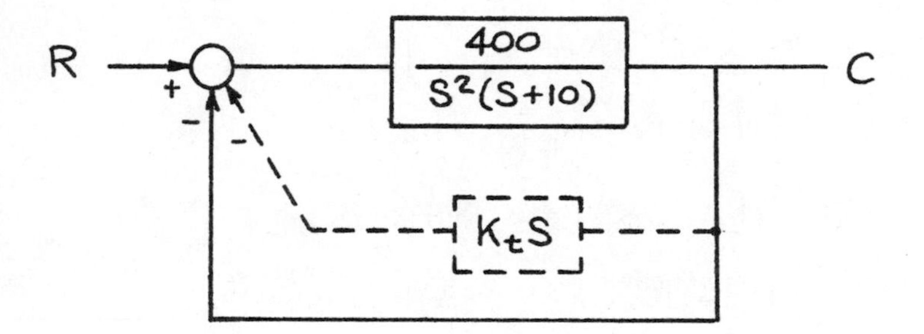

REQUIRED:

(a) Find the minimum value of K_t such that the system will just be stable.

(b) If the value of K_t (found in Part (a) were increased by a factor of 1.25 [i.e., 1.25 x K_t min], determine the

approximate step response characteristics.
(Hint: one method of an approximate solution is to use only a reasonably accurate root-locus sketch and then approximate with "standardized" 2nd order curves.)

Solution

(a)

For stability use Routh-Hurwitz criterion for the characteristic polynominal:

$$G_{SYST} = \frac{C}{R} = \frac{400}{S^2(S+10) + (1+K_tS)400}$$

Since equivalent block diagram may be given as:

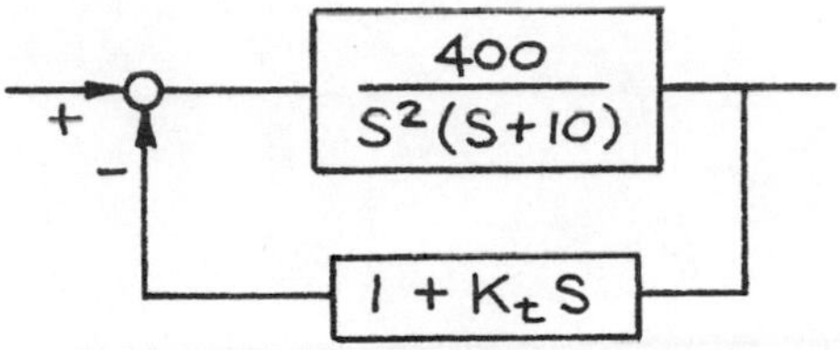

Characteristic Polynomial $= S^3 + 10S^2 + 400K_tS + 400$

Routh-Hurwitz array:

S^3	1	$400K_t$
S^2	10	400
S^1	X_1	

$$X_1 = \frac{(10)(400K_t) - (1)(400)}{10}$$

$\therefore K_t > 0.1$ for $X_1 > 0$

(K_t minimum $= 0.1$)

(b)

Plot root-locus for HG:

Let $H = 1 + K_tS = 1 + 0.125S = 0.125(8+S)$

$$\therefore HG = \frac{(0.125)(400)(S+8)}{S^2(S+10)} = \frac{50(S+8)}{S^2(S+10)}$$

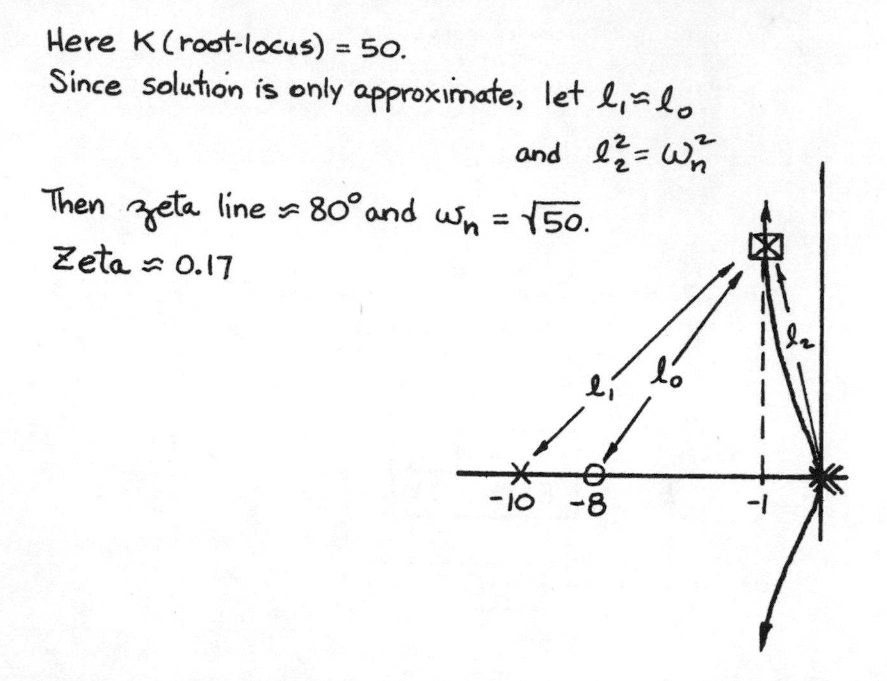

Here K (root-locus) = 50.

Since solution is only approximate, let $l_1 \approx l_0$ and $l_2^2 = \omega_n^2$

Then zeta line $\approx 80^\circ$ and $\omega_n = \sqrt{50}$.

Zeta ≈ 0.17

From any standard 2nd order transient response curves, one can easily determine percent overshoot and time to first peak, t_p.

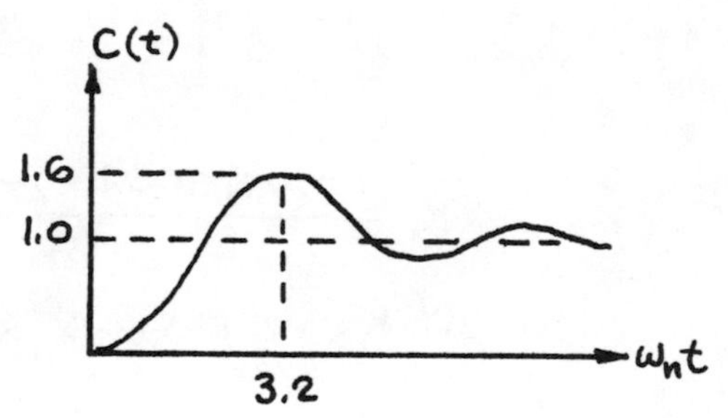

% O.S. = 60%

$$t_p = \frac{3.2}{\omega_n} = 0.45 \text{ seconds}$$

PROBLEM 3. Electronics. (A typical question involves the small signal transistor or vacuum tube equivalent circuit amplifier - usually one stage.)

The following questions apply to the circuits 1 and 2 and their common graphical operation characteristic diagram shown on the following page.

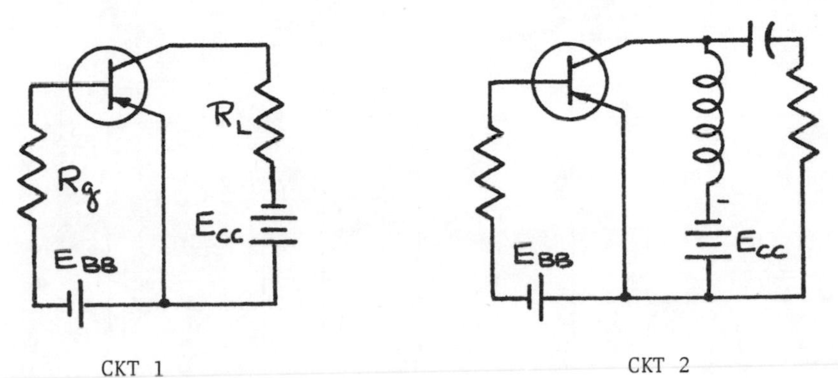

CKT 1 CKT 2

REQUIRED:

(a) What are the types and classes (mode) of amplifiers shown?

(b) What are their power outputs?

(c) What are their transistor-power dissipation loads during no-signal and maximum signal conditions?

(d) What are their collector efficiencies while under maximum conditions?

(e) What are their per cent of second harmonic distortion under maximum conditions?

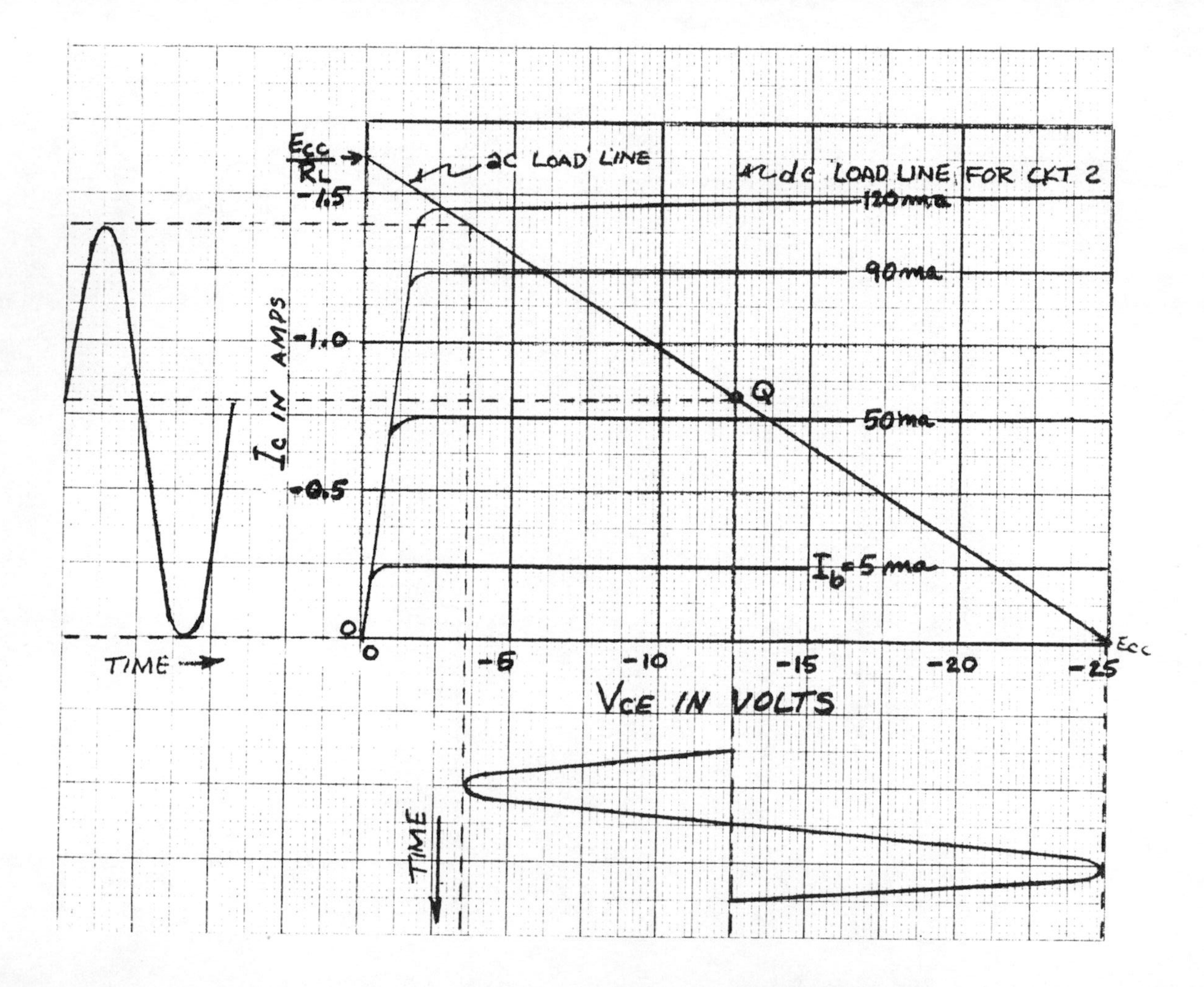
E_{CC}/R_L
ac LOAD LINE
dc LOAD LINE FOR CKT 2
-1.5
120 ma
90 ma
-1.0
I_C IN AMPS
Q
50 ma
-0.5
I_b=5 ma
0
TIME
0
-5
-10
-15
-20
-25
E_{CC}
V_{CE} IN VOLTS
TIME

Solution

(a) Common emitter, Class A amplifiers because the collector current flows during the entire cycle.

(b) Signal power output,

$$P_o = \frac{(v_{CE,max} - v_{CE,min})}{2\sqrt{2}} \times \frac{(i_{C,max} - i_{C,min})}{2\sqrt{2}}$$

$$= 3.76 \text{ watts}$$

The signal power output is the same for both circuits.

(c) No signal condition:

Q point given to be at $V_{CQ} = -12.5$ V, $I_{CQ} = -0.8$ A for both circuits.

$\therefore$ Power dissipation in transistor $= 12.5 \times 0.8 = 10$ watts for both circuits.

Maximum signal condition:

(i) Circuit 1:

Average power dissipated in collector

$$= \left(\frac{v_{CE,max} + v_{CE,min}}{2}\right)\left(\frac{i_{C,max} + i_{C,min}}{2}\right) - \frac{(v_{CE,max} - v_{CE,min})}{2\sqrt{2}} \times \frac{(i_{C,max} - i_{C,min})}{2\sqrt{2}}$$

$$= 6.2 \text{ watts}$$

(ii) Circuit 2:

Average power dissipated in transistor

$$= \left(\frac{V_{CE,max} + V_{CE,min}}{2}\right)\left(\frac{i_{C,max} + i_{C,min}}{2}\right)$$

$$- \frac{(V_{CE,max} - V_{CE,min})}{2\sqrt{2}} \times \frac{(i_{C,max} - i_{C,min})}{2\sqrt{2}}$$

$$= 9.975 \text{ watts} - 3.76 \text{ watts}$$

$$\simeq 6.2 \text{ watts}$$

(d) Collector efficiency

Circuit 1: P_o = ac power output to load = 3.76 watts

P_i = Power input = $V_{CC}\ I_{C,average}$

= 25 x 0.7 = 17.5 watts

$$\therefore \text{Collector Efficiency, } \eta_c = \frac{3.76}{17.5} \times 100 = 21.5\,\%$$

Circuit 2: AC power output is the same, P_o = 3.76 watts

Power input is reduced.

$$V_{CC} = \frac{V_{CE,max} + V_{CE,min}}{2} = 14.25 \text{ volts}$$

$$\therefore \text{Power input, } P_i = V_{CC}\, I_{C,average}$$

$$= 14.25 \times 0.7 = 9.975 \text{ watts}$$

$$\therefore \text{Collector Efficiency, } \eta_c = \frac{3.76}{9.975} \times 100 = 37.7\%$$

The collector efficiency is increased. This increase in efficiency is due to the fact that there is no dc power dissipated in the load.

(e)

Let $i_c(t) = I_o + I_1 \cos \omega t + I_2 \cos 2\omega t$

at $t=0$, $i_c = i_{c,max} = I_o + I_1 + I_2$ (1)

at $\omega t = \frac{\pi}{2}$, $i_c = I_{CQ} = I_o - I_2$ (2)

at $\omega t = \pi$, $i_c = i_{c,min} = I_o - I_1 + I_2$ (3)

From Eqns (1) and (3)

$$I_1 = \frac{i_{c,max} - i_{c,min}}{2} \quad (4)$$

From Eqns (2) and (3)

$$I_2 = \frac{i_{c,min} - I_{CQ} + I_1}{2}$$

$$= \frac{i_{c,min} - I_{CQ} + \frac{i_{c,max}}{2} - \frac{i_{c,min}}{2}}{2}$$

$$= \frac{i_{c,max} + i_{c,min} - 2I_{CQ}}{4} \quad (5)$$

$\therefore$ Percentage Second Harmonic Distortion

$$= \left|\frac{I_2}{I_1}\right| \times 100 = \frac{|-1.4 + 0 + 1.6|}{2\,|(-1.4 - 0)|} \times 100$$

$= 7.1\%$

PROBLEM 4. Power & Systems. (At least one problem is usually included that involves power factor correction - either by capacitors or using a synchronous motor.)

Originally it is planned to furnish a plant load requirement of 1000 H.P. at 2200 volt, 3-phase, by induction motors operating at 80% power factor and 90% efficiency.

REQUIRED:

(a) Find the line current necessary to supply this load and generator capacity.

(b) Assume that rather than supplying the 1000 H.P. by induction motors, it is decided to produce 400 H.P. of this load by a synchronous motor operating in the over-excited leading mode of 85% (assume same efficiency as for an induction motor). Find the new total line current requirement and the overall power factor.

(c) If rather than installing the 400 H.P. synchronous motor (in Part b) it is considered feasible to use power factor correcting capacitors for the 1000 H.P. motors in Part (a) to achieve the same power factor correcting as obtained in Part (b). Determine the size (KVAR) of the capacitors needed.

Solution

(a) Power input to motors:

$$\frac{1000}{0.9} = 1111 \text{ HP} \equiv 830 \text{ KW}$$

Generator capacity

$$\frac{830 \text{ KW}}{0.8} = 1036 \text{ K.V.A.}$$

Current requirement

$$\frac{1036 \times 1000}{\sqrt{3} \times 2200} = 272 \text{ A}$$

(b) Power requirements:

Induct. motor input

$$\frac{600 \text{ HP}}{0.9} \times \frac{746}{1000} = 497 \text{ KW}$$

Induct. motor current

$$\frac{497 \times 1000}{\sqrt{3} \times 2200 \times 0.8} = 163 \text{ A.}$$

Synch. motor input

$$\frac{400 \text{ HP}}{0.9} \times \frac{746}{1000} = 332 \text{ KW}$$

Synch. motor current

$$\frac{332 \times 1000}{\sqrt{3} \times 2200 \times 0.85} = 102 \text{ A.}$$

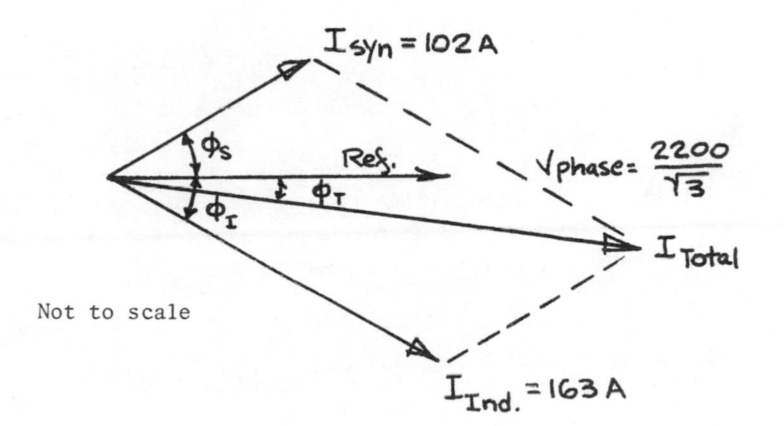

Not to scale

$$I_{Total} = \sqrt{(I_I \cos\phi_I + I_S \cos\phi_S)^2 + (I_I \sin\phi_I - I_S \sin\phi_S)^2}$$

$$= \sqrt{(163 \times 0.8 + 102 \times 0.85)^2 + (163 \times 0.6 - 102 \times 0.53)^2}$$

$$= 221 \text{ A } \angle\phi_T, \quad \phi_T = 11.5° \text{ lagging}$$

(c)

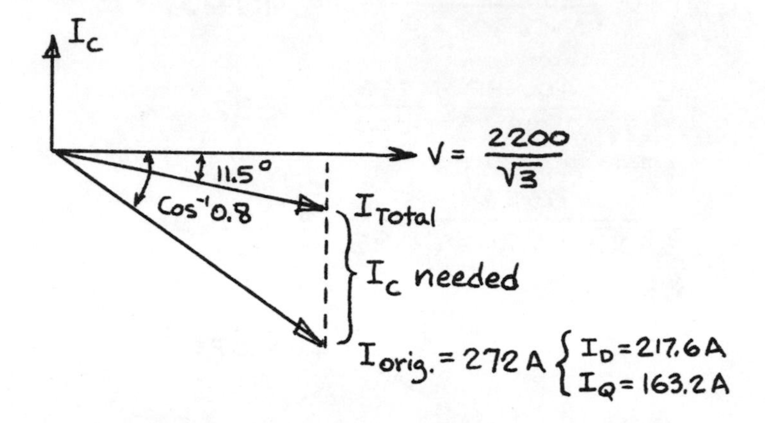

$$I_{Total} = \frac{I_D}{\cos 11.5^\circ} = \frac{217.6}{0.98} = 222.06$$

$$I_{Total_Q} = 222.06 \sin 11.5^\circ = 44.3 \text{ A}$$

$$\therefore I_C = I_{O_Q} - I_{T_Q} = 163.2 - 44.3 = 118.9 \text{ A}$$

$$\therefore KVAR = \frac{118.9 \times 2200}{1000\sqrt{3}} = 151 \text{ KVAR/Phase}$$

PROBLEM 5. Measurement. (Here problems may be from a bridge circuit using an a.c. source and converting to d.c. for a D'Arsonval meter
- or perhaps temperature measurement;
and, usually some form of wave analysis.)

A temperature-control system is being used in a chemical process. The temperature is controlled by a heater (and is assumed to be uniform throughout).

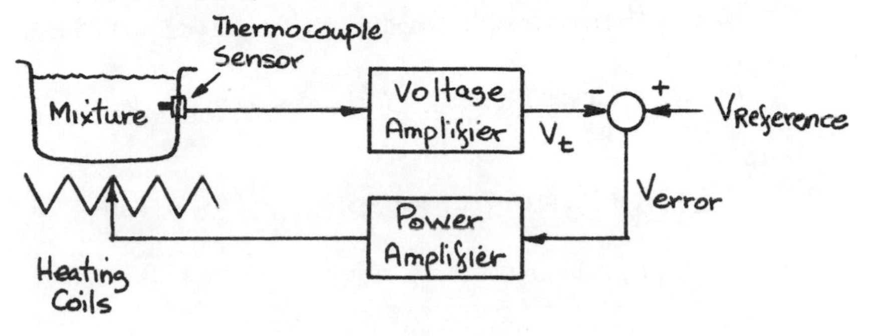

Assume the heater raises the mixture 10°C per K.W. of heating power and the power amplifier has a gain of 500 KW/volt.

The thermocouple sensor produces 0.003 m.v./°C and the voltage amplifier gain is 1000.

REQUIRED:

(a) Find the value of V_R if the mixture temperature should be 90°C. You may assume the ambient temperature is 25°C.

(b) Assume the power amplifier produces a full wave rectified voltage (60 hz) of 10 volts peak for it's output per 0.1 volt d.c. on it's input. Determine the necessary heating coil resistance so that the heating coil-power amplifier combination may be rated at 500 KW/volt.

Solution

(a) Heater power required = $\frac{90° - 25°}{10°/KW}$ = 7.5 KW

$$\therefore \text{amplifier input, } V_e = \frac{7.5 \text{ KW}}{500 \text{ KW/volt}} = 15 \times 10^{-3} \text{ volts}$$

at 90°C thermocouple produces (90°)(0.003) = 0.27 m.v.

$$\text{or } V_t = (1000)(0.27 \times 10^{-3}) = 0.270 \text{ volts}$$

Thus

$$V_R = V_t + V_e = 0.270 + 0.015 = 0.285 \text{ volts}$$

(b) 1 V. E_{in}(dc) produces 100 volts (peak) in output or 70.7 effective rms volts.

And, since the voltage "works into" a pure resistance heating coil,

$$P = E^2/R$$

Therefore, from 70.7 volts we need to produce 500 KW

$$500{,}000 = \frac{E^2}{R} = \frac{(70.7)^2}{R}$$

$$R = \frac{5000}{500{,}000} = 0.01\ \Omega$$

PROBLEM 6. **Illumination**. (This particular problem is from the Illumination chapter. It is felt to be representative of the National Electrical Engineering exam questions.)

The following problem concerns lighting design, and answers should be based on the latest editions of applicable references:

(a) Give the equation for calculating the cavity ratios, and name the ratios required for a room illuminated with pendent fixtures.

(b) What two values of reflections are adjusted by use of the foregoing ratios?

(c) A maintenance factor is normally used in interior lighting calculations. Give three main reasons why. (Neglect wiring)

(d) A classroom for a high school 24 ft x 32 ft x 10 ft high is to be illuminated, using recessed florescent fixtures with low brightness acrylic prismatic lenses and two 40-watt RS (3100 Lu) lamps per fixture. Compute the number of fixtures required assuming a CU of 0.6.

What would be an appropriate M.F.? What is a recommended footcandle level? How many fixtures are required theoretically?

(e) Sketch the actual layout showing the fixture spacing and distance to walls, for the room in Part (d). Assume the front of the room is at the 24-ft end.

What is the computed maintained footcandle level of the layout shown?

Refer to Illumination 5 in Chapter 8 for the solution.

PROBLEM 7. Communications. (One might expect a problem on transmission lines, use of the Smith chart, field and waves, or a full communication systems problem. The actual problem presented here is Problem 5 from the Communications chapter.)

In the modulating circuit sketched below, the modulating signal (see spectrum sketch) is limited to angular frequencies where

$$\omega_{m_1} < \omega < \omega_{m_2} \lll \omega_c.$$

where ω_m = modulating frequency

ω_c = carrier frequency

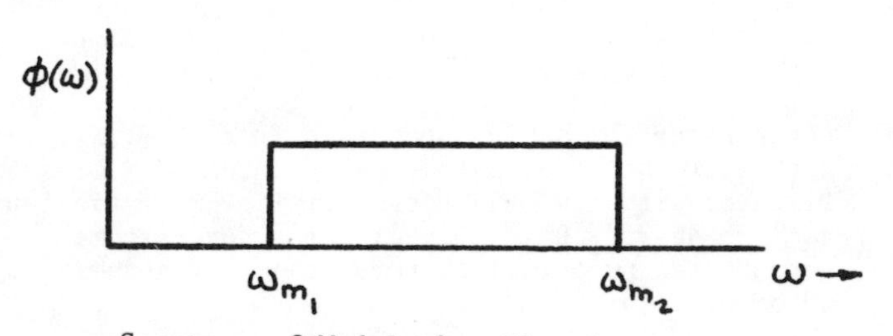

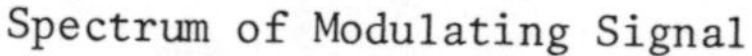
Spectrum of Modulating Signal

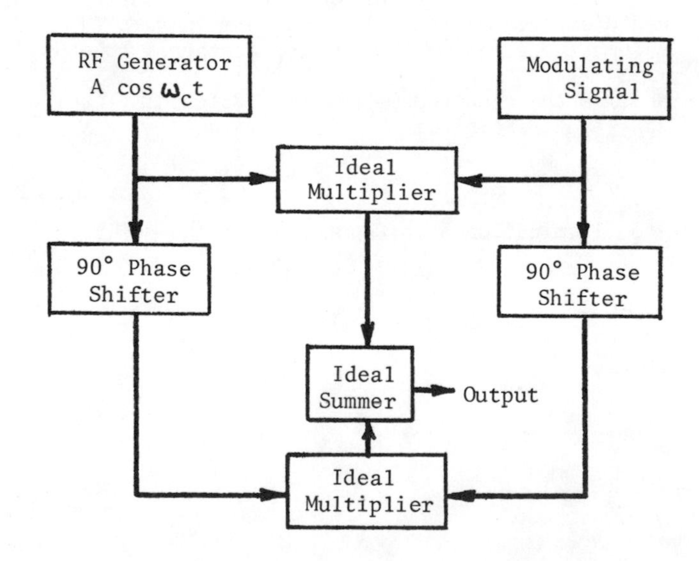

REQUIRED:

(a) Sketch the spectrum of the output.

(b) What is this type of modulated signal called?

(c) Sketch another circuit which would produce the same output spectrum.

Refer to Chapter 6 (Electronics & Communications), Problem 5 for the solution.

PROBLEM 8. Circuits. (Another frequently asked circuits problem is one involving filter theory, either constant "K", "M" derived, or similar to the following.)

Consider the following filter driven by a current source and the output goes to an infinite impedance load. [Note: C_L is only in the circuit for Part (b).]

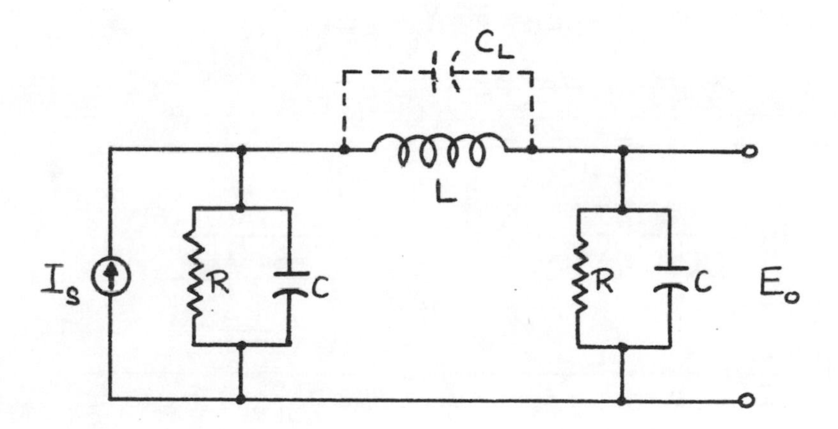

REQUIRED:

(a) Find the transfer function relating the output voltage to the input current, E_o/I_s. From your solution describe the behavior of the circuit (i.e., high pass, low pass, maximally flat or what? Why?).

(b) If a capacitor, C_L, is placed across L, describe the behavior of the filter in terms of any specific important frequency.

Let L = 0.01 h, C_L = C = 0.01 μf, R = 100 Ω

Solution

(a)

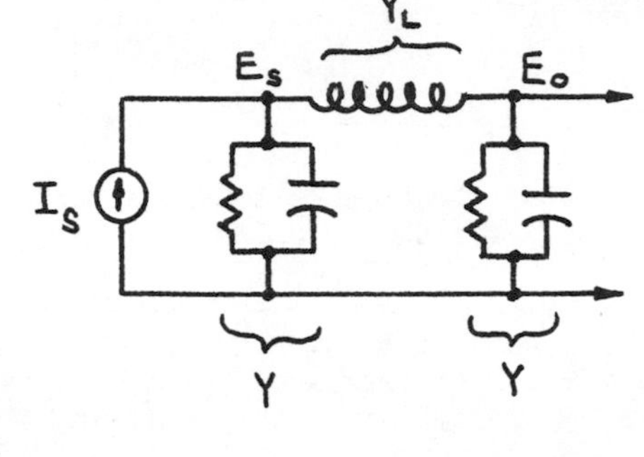

$$Y = \frac{1}{R} + SC = G + SC \qquad Y_L = \frac{1}{SL}$$

Σi's at E_s: $\quad I_s = E_s Y + (E_s - E_o) Y_L$

at E_o: $\quad (E_s - E_o) Y_L = E_o Y$

Eliminating E_s gives:

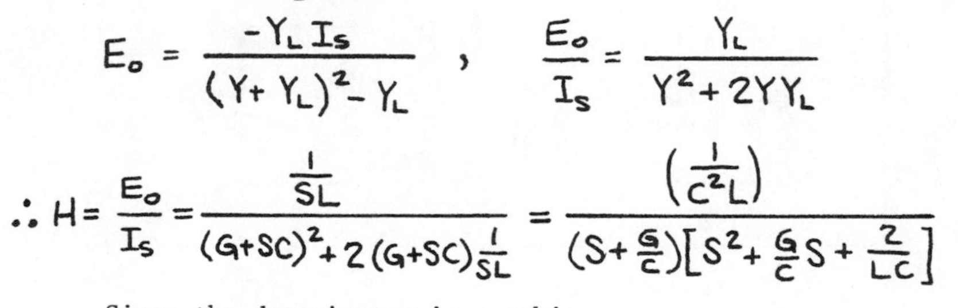

$$E_o = \frac{-Y_L I_s}{(Y+Y_L)^2 - Y_L}, \qquad \frac{E_o}{I_s} = \frac{Y_L}{Y^2 + 2YY_L}$$

$$\therefore H = \frac{E_o}{I_s} = \frac{\frac{1}{SL}}{(G+SC)^2 + 2(G+SC)\frac{1}{SL}} = \frac{\left(\frac{1}{C^2L}\right)}{\left(S+\frac{G}{C}\right)\left[S^2 + \frac{G}{C}S + \frac{2}{LC}\right]}$$

Since the denominator is a cubic, two of the roots may be complex and one must be real. The pole locations turn out to be on a semi-circle and can be adjusted to 60° apart (the maximum separation the complex poles may have without peaks appearing in the output); this would yield a <u>maximally flat</u> low-pass filter design.

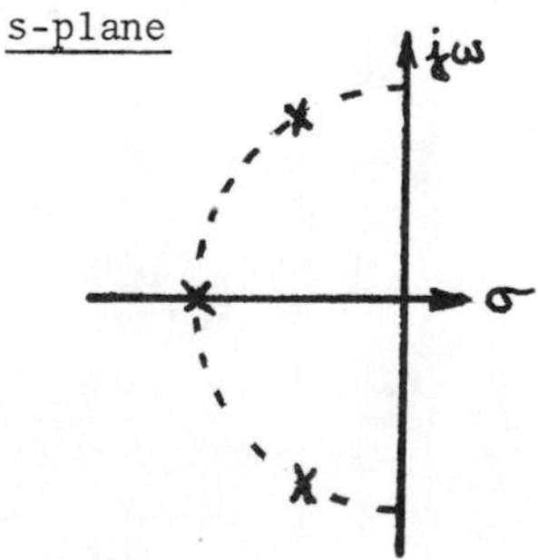

(b)

C_L

L

$$Y_L' = \frac{1}{SL} + SC_L$$

$$\therefore \frac{E_o}{I_s} = H = \frac{SC_L + \frac{1}{SL}}{(G+SC)\left[G+SC+2SC_L+\frac{2}{SL}\right]} \qquad \text{Let } C_x = C + 2C_L$$

$$= \left(\frac{C_L}{CC_x}\right)\left[\frac{\left(S^2 + \frac{1}{LC_L}\right)}{\left(S+\frac{G}{C}\right)\left(S^2 + \frac{G}{C_x}S + \frac{2}{LC_x}\right)}\right]$$

The form of the denominator is the same as previously found (except that one can't get a maximally flat design); however, the numerator gives two complex zeros on the imaginary axis:

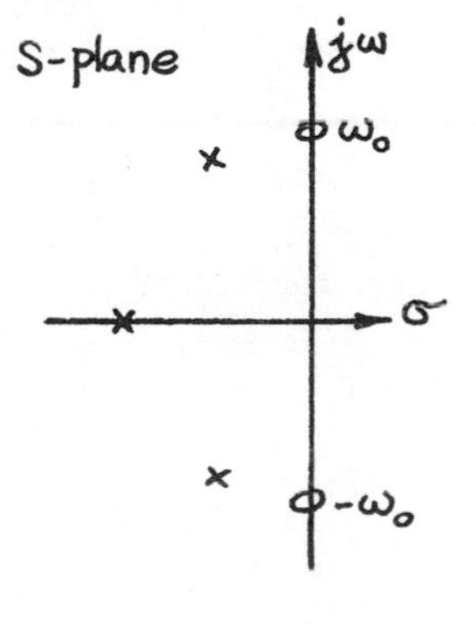

$$\omega_o = \sqrt{\frac{1}{LC_L}}$$

$$= \sqrt{\frac{1}{(10^{-2})(10^{-8})}}$$

$$= 10^5 \text{ rad/sec} \equiv 16 \text{ Khz}$$

Thus the 16 Khz signal will not pass.

PROBLEM 9. Economics. [The morning session contains one (and only one) economic analysis problem. Because the same problem is used for all branches of engineering, no special E.E. knowledge is required in its solution. Below is a rather typical problem.]

A trust fund is to be established to

(a) Provide $750,000 for the construction and $250,000 for the initial equipment of an Electrical Engineering Laboratory.

(b) Pay the laboratory annual operating costs of $150,000 per year.

(c) Pay for $100,000 of replacement equipment every four years, beginning four years from now.

At 4% interest, how much money would be required in the trust fund to provide for the laboratory and equipment and its perpetual operation and equipment replacement?

Solution: For the solution see Economics 9 in Chapter 9.

PROBLEM 10. Machines. (This area seems "wide open" as questions can be transformer connections, such as 3-phase zig-zag voltage ratios and loading; transformer efficiencies requiring the use of equivalent circuits from open circuit - short circuit data; complete d.c. motor analysis involving starting torques and efficiency calculations; a.c. machines and loading.)

A 12,500 K.V.A., 6600 volt, 3600 r.p.m., 60 hz, three-phase, Y connected alternator has magnetization and short circuit characteristic curves as shown on the next page.

REQUIRED: Determine the percentage voltage regulation for a 0.707 lagging power factor. Let the a.c. armature resistance be 0.5 ohms and make (and state) any reasonable assumption necessary for your solution.

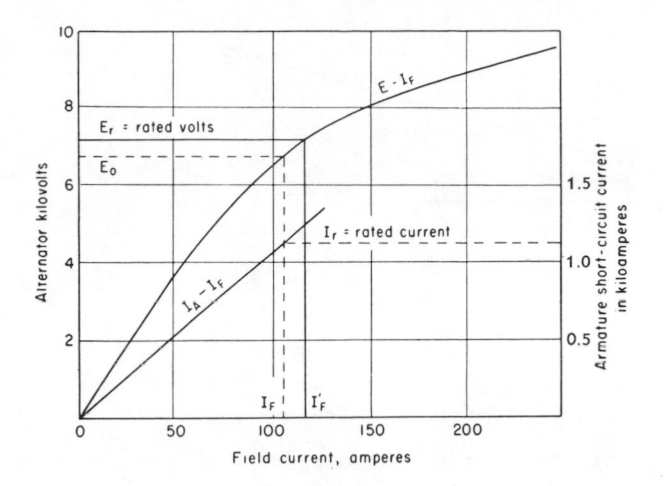

Solution

$$\text{Rated } I = \frac{12{,}500{,}000}{6600\sqrt{3}} = 1090 \text{ A}$$

Field current necessary to give short circuit line current, from graph:

$$I_f \text{ (d.c.)} = 105 \text{ A}$$

Terminal voltage (per phase) from graph:

$$\frac{6300}{\sqrt{3}} = 3630 \text{ volts}$$

Terminal rated voltage (per phase)

$$\frac{6600}{\sqrt{3}} = 3800 \text{ volts}$$

(Here, one could find the synchronous reactance or impedance by taking these operating values and defining

$$X_S \approx \frac{3630 \text{ V}}{1090 \text{ A}},$$

however, since only one condition is asked for, the synchronous voltage drop is the 3630 volts.)

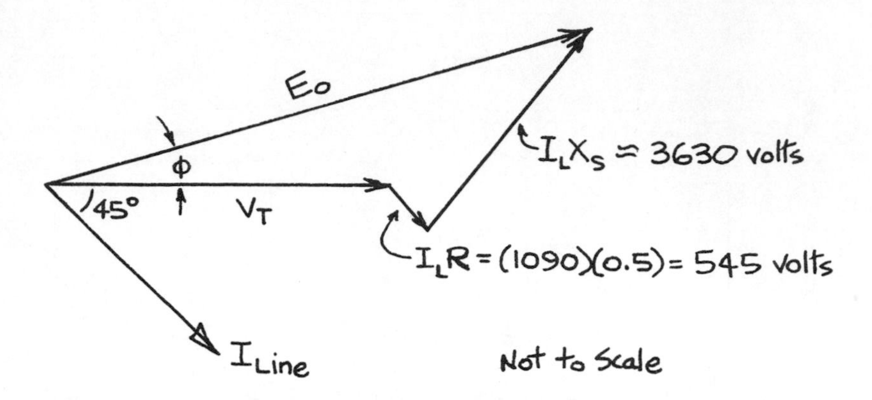

$$E_o = V_T + R_A I_L + jX_S I_L$$
$$= 3800 + 545(\cos 45° - j\sin 45°) + 3630(\cos 45° + j\sin 45°)$$
$$= 6752.4 + j\,2181.6 = 7096 \angle\phi \text{ volts}$$

$$\therefore \ \% \text{ V.R.} = \frac{E_o - V_T}{V_T}(100) = \frac{7096 - 3800}{3800}(100)$$
$$= 86.7\%$$

Note that the calculated no-load voltage is

$$7096\sqrt{3} = 12,290 \text{ volts}$$

which is well beyond the range of available field current; thus the saturated limit of no-load voltage would be between 9,000 and 10,000 volts. Thus the saturated value of synchronous reactance would be less than the value used here.

AFTERNOON SESSION

The afternoon session is also a group of ten problems and again the task has been to work any four of them. These problems are essentially the same degree of difficulty as the morning session.

A typical format:

1. Economics
2. Illumination
3. Communications - Gaussian Distribution
4. Measurement - Using operational amplifiers
5. Machines - d.c. motor loading
6. Power & Systems - Harmonic shunting filter
7. Circuits - Low pass and band pass filters
8. Control Systems - Stability problem
9. Power & Systems - Motor loading and wire size
10. Machines - Transformer equivalent circuit

APPENDIX

This appendix represents the authors review as to the nature and classification of National Electrical Engineering Examination problems over the last several years. The classification should not be rigorously interpreted as being anything more than a guide. Future examinations may follow these patterns, but from here on, one may only speculate.

EXAM 1

Problem

1. Economics. Compute uniform annual amount equivalent to an irregular series of money amounts.
2. Circuits. a.c. equivalent circuits.
3. Electronics. Triode power rating and load line.
4. Circuits. a.c. circuits.
5. Electronics. Transistor logic circuit and truth table.
6. Circuits. a.c. loop currents.
7. Power & Systems. Power factor correction - synchronous motors.
8. Power & Systems. Power factor correction - capacitors.
9. Control Systems. Transfer function from Bode Plot.
10. Electronics. Coupled transistor stages.

Afternoon

Problem

1. Economics. Two alternatives, one with perpetual life. Fixed output. Compute uniform equivalent annual cost (EUAC).
2. Electronics. Single stage transistor amplifier analysis.
3. Machines. 3-phase induction motor analysis.
4. Measurement. Voltage regulation using nonlinear resistance.
5. Power & Systems. Open delta transformer loading.
6. Electronics. 3-stage transistor amplifier bandwidth analysis.
7. Control Systems. Steady-state error & specifications for system.
8. Electronics. 1-stage vacuum tube, transformer coupled, analysis.
9. Illumination. Power requirements using fluorescent lamps.
10. Circuits. R.L.C. pulsed analysis.

EXAM 2

Problem

1. Economics. Present worth of a bond.
2. Power & Systems. Electric field for high voltage insulator.
3. Power & Systems. Power factor correction - synch. impedance.
4. Power & Systems. Power factor correction - use of National Elect. Code.
5. Electronics. Differential amplifier using vacuum triodes.
6. Communications. Harmonic analysis for telephone circuit.
7. Electronics. 1 stage transistor biasing analysis.
8. Electronics. 1 stage power vacuum tube analysis.
9. Control System. Steady state error and stability analysis.
10. Circuits. RLC parallel transient analysis.

Afternoon

Problem

1. Economics. Two alternative rent or buy question.
2. Circuits. RLC transient.
3. Electronics. 1 stage transistor amplifier analysis.
4. Machines. Autotransformer rating.
5. Power & Systems. Power factor correcting using capacitors.
6. Machines. Induction motor equivalent circuit analysis.
7. Control Systems. Lead network compensation design.
8. Communications. Field analysis for calculating insulation.
9. Electronics. 1 stage transistor analysis.
10. Measurement. Field winding protection using diodes.

EXAM 3

Problem

1. Economics. Annual cost of "Energy Peaking Capacity".
2. Control Systems. Root locus plot.
3. Machines. Use of inverter to drive motors.
4. Power & Systems. 3 phase transmission line
5. Measurement. Analysis of phase shift circuit.
6. Circuits. RLC transient driven by a.c. source.
7. Electronics. 2 stage transistor amplifier linear analysis.
8. Electronics. Transistor equivalent circuit analogy.
9. Economics. Unit cost for various production lot sizes.
10. Machines. Design of d.c. motor starter circuit.

Afternoon

Problem

1. Economics. Optimum allocation of load to minimize generating plant fuel cost.
2. Illumination. Mercury vapor vs. incandescent lighting.
3. Electronics. Vacuum tube cathode-follower analysis.
4. Machines. d.c. motor equivalent circuit calculation.
5. Power & Systems. Total power calculation from various power-factor loads.
6. Electronics. Transistor amplifier bandwidth calculation.
7. Measurements. Wave shaping circuit using diodes.
8. Power & Systems. Fault calculations and sequence networks.
9. Circuits. Filter network calculation.
10. Economics. Three alternatives. Neither input nor output fixed.

EXAM 4

Problem

1. Economics. Breakeven analysis by method of equivalent annual cost.
2. Power & Systems. Power factor correction using synchronous motors.
3. Circuits. RLC "M-coupled" transient problem.
4. Measurement. Wave form analysis from diode clipped sinusoids.
5. Power & Systems. 3 phase transmission line calculation.
6. Power & Systems. Power output from hydro-electric plant.
7. Control Systems. Range of stability for non-unity feedback system.
8. Power & Systems. Unbalanced 3 phase power measurement.
9. Measurement. Power output from SCR circuit.
10. Communications. Standing wave ratio from Klystron for matching network.

Afternoon

Problem

1. Economics. 2 alternatives. Compute expected present worth of cost.
2. Illumination. Mercury vapor calculations utilizing "isolux curves."
3. Measurement. Optical sensor sensitivity and response time.
4. Measurement. Power supply regulation using zener diodes.
5. Electronics. Biasing a transistor stage.
6. Circuits. RLC - differential equation to Laplace Transformation.
7. Measurements. Steady state phase shifting network analysis.
8. Power & Systems. Power factor correction - 3 phase.
9. Circuits. Matrix form of network equation.
10. Power & Systems. 3 phase transmission line analysis.

EXAM 5

Problem

1. Economics. Two alternatives. Neither input nor output fixed.
2. Circuits. Series RLC resonance and "bandwidth" analysis.
3. Machines. Induction motor starting analysis.
4. Machines. Transformer connections and loading for 3 phase.
5. Power & Systems. 3-phase transmission line "voltage regulation."
6. Power & Systems. Loads and wire size using National Electrical Code.
7. Circuits. Finding Thevenin's equivalent circuit for resistive network.
8. Control Systems. Stability and error analysis.
9. Communications. Transmission line, S.W.R.
10. Electronics. Transformer coupled transistor amplifier.

Afternoon

Problem

1. Economics. Two alternatives. Neither input nor output fixed.
2. Power & Systems. Loading of a 3-phase system
3. Power & Systems. Transformer loading of a 3-phase system.
4. Power & Systems. 3-phase transmission line analysis.
5. Power & Systems. 3-phase power system loading.
6. Power & Systems. Fault calculations - high voltage line.
7. Circuits. Multiloop resistive network analysis.
8. Circuits. Maximum power transfer for an a.c. circuit.
9. Communications. Frequency response for cascaded amplifier stages.
10. Electronics. Tuned transformer-coupled transistor amplifier.

EXAM 6

Problem

1. Economics. Single stage or two stage construction.
2. Power & Systems. Conductor sizes for given 3-phase loads.
3. Power & Systems. 3-phase power factor correction.
4. Power & Systems. Power line connections, loads, and spacing.
5. Electronics. Z_o calculation for transistor configuration.
6. Measurement. Temperature sensing with balancing bridge circuit.
7. Circuits. Current-power relationships for a capacitive circuit.
8. Machines. d.c. shunt motor analysis.
9. Control Systems. Stability analysis for unity feedback system.
10. Electronics. Equivalent circuit for "black box" amplifier.

Afternoon

Problem

1. Economics. Present worth and Present worth of perpetual service (capitalized cost).
2. Electronics. Transistor amplifier using "diode biasing."
3. Power & Systems. "Short Circuit" circuit breaker rating for given transformer.
4. Circuits. Steady state RLC a.c. circuit analysis.
5. Illumination. Plant illumination calculations for fluorescent lamps.
6. Machines. Transformer equivalent circuit calculations.
7. Circuits. RC circuit analysis of stored energy.
8. Electronics. Pair of FET's for transistor amplifier analysis.
9. Electronics. Logic analysis of I.C.'s for NAND gates.
10. Power & Systems. Line "bus loading" by induction motors.